T0235771

# OFDM Systems
# for Wireless Communications

# Synthesis Lectures on Algorithms and Software in Engineering

Editor
**Andreas S. Spanias,** *Arizona State University*

**OFDM Systems for Wireless Communications**
**Adarsh B. Narasimhamurthy, Mahesh K. Banavar, and Cihan Tepedelenlioğlu**
**2010**

**Algorithms and Software for Predictive Coding of Speech**
**Atti Venkatraman**
**2010**

**MATLAB Software for the Code Excited Linear Prediction Algorithm: The Federal Standard–1016**
**Karthikeyan N. Ramamurthy and Andreas S. Spanias**
**2010**

**Advances in Waveform-Agile Sensing for Tracking**
**Sandeep Prasad Sira, Antonia Papandreou-Suppappola, and Darryl Morrell**
**2008**

**Despeckle Filtering Algorithms and Software for Ultrasound Imaging**
**Christos P. Loizou and Constantinos S. Pattichis**
**2008**

OFDM Systems for Wireless Communications

Adarsh B. Narasimhamurthy, Mahesh K. Banavar, and Cihan Tepedelenlioğlu

ISBN: 978-3-031-00385-1     paperback
ISBN: 978-3-031-01513-7     ebook

DOI 10.1007/978-3-031-01513-7

A Publication in the Springer series
*SYNTHESIS LECTURES ON ALGORITHMS AND SOFTWARE IN ENGINEERING*

Lecture #5
Series Editor: Andreas S. Spanias, *Arizona State University*
Series ISSN
Synthesis Lectures on Algorithms and Software in Engineering
Print 1938-1727    Electronic 1938-1735

# OFDM Systems
# for Wireless Communications

Adarsh B. Narasimhamurthy, Mahesh K. Banavar, and Cihan Tepedelenlioğlu
Arizona State University

*SYNTHESIS LECTURES ON ALGORITHMS AND SOFTWARE IN ENGINEERING #5*

# ABSTRACT

Orthogonal Frequency Division Multiplexing (OFDM) systems are widely used in the standards for digital audio/video broadcasting, WiFi and WiMax. Being a frequency-domain approach to communications, OFDM has important advantages in dealing with the frequency-selective nature of high data rate wireless communication channels. As the needs for operating with higher data rates become more pressing, OFDM systems have emerged as an effective physical-layer solution.

This short monograph is intended as a tutorial which highlights the deleterious aspects of the wireless channel and presents why OFDM is a good choice as a modulation that can transmit at high data rates. The system-level approach we shall pursue will also point out the disadvantages of OFDM systems especially in the context of peak to average ratio, and carrier frequency synchronization. Finally, simulation of OFDM systems will be given due prominence. Simple MATLAB programs are provided for bit error rate simulation using a discrete-time OFDM representation. Software is also provided to simulate the effects of inter-block-interference, inter-carrier-interference and signal clipping on the error rate performance. Different components of the OFDM system are described, and detailed implementation notes are provided for the programs. The program can be downloaded from `http://www.morganclaypool.com/page/ofdm`

# KEYWORDS

multi-carrier, orthogonal frequency division multiplexing (OFDM), frequency domain, carrier frequency offset, peak-to-average power ratio, simulations

# Contents

# Preface

Orthogonal Frequency Division Multiplexing (OFDM) is a multicarrier communication scheme widely adopted in the wireless communications industry. In this book, we provide a brief and comprehensive coverage of the OFDM system model, an overview of its advantages and disadvantages, along with MATLAB codes for simulation. This book is intended for practitioners or students with some elementary knowledge of digital communications. The main focus of this book is to aid readers in understanding the workings of a point to point baseband OFDM system and understanding how to simulate performance under certain impairments. A unique feature of the book is its emphasis on discrete-time representations which are used to simulate OFDM systems. In order to make the book accessible to a wider audience, we present several simulations, which provide a deeper insight into the subject. An extensive list of references is also included to support further reading.

We begin by highlighting the benefits that OFDM offers over the conventional frequency division multiplexing scheme in terms of bandwidth efficiency and implementation complexity. Following this, we motivate the need for OFDM systems by providing a brief introduction to wireless fading channels, with special emphasis on the time varying and frequency selective nature of such channels. We demonstrate that complex equalization at the receiver, which would be required for communication over frequency selective channels, are not needed in the case of OFDM systems, further motivating its use. Different variations on the basic OFDM system are also presented to illustrate its versatility. Drawbacks of OFDM such as high peak-to-average power ratio (PAPR) at the transmitter, and carrier frequency offset (CFO) at the receiver are described, along with their adverse effects on system performance. Techniques to mitigate their effects are also presented. All these concepts are supported with simulations. The programs used for these simulations, with detailed comments, are also provided.

We would like to thank Professor Andreas Spanias, for providing us with the opportunity to author this book, and Morgan & Claypool publishers for working with us in producing this manuscript.

Adarsh B. Narasimhamurthy, Mahesh K. Banavar, and Cihan Tepedelenlioğlu
February 2010

# CHAPTER 1

# Introduction

Next-generation wireless communication systems mandate data rate intensive applications like multimedia services, data transfer, audio, streaming video, leading to future wireless terminals being capable of connecting to various networks to support services like switched traffic, IP data packets and broadband streaming services. Additionally, with the growth of Internet applications and wireless users, many wireless local area network (WLAN) standards, including IEEE802.11, permit mobile connectivity to the Internet. With a surging demand for wireless Internet connectivity, new WLAN standards have been developed including IEEE802.11b, popularly known as Wi-Fi, that provides up to 11 Mb/s raw data rate, and more recently IEEE802.11g that provides wireless connectivity with speeds up to 54 Mb/s. High data rates are a requirement for not only wireless networks but also in broadcasting standards like Digital Audio Broadcast (DAB) [1], Digital Video Broadcasting-Terrestrial (DVB-T) [2] and the HiperLAN-2 standards in Europe, the Integrated Services Digital Broadcasting (ISDB) in Japan and the Korean Digital Multimedia Broadcasting-Terrestrial (DMB-T) standard [3]. As a solution to their requirements for high data rates, all these standards use multicarrier communications, and in most cases, Orthogonal Frequency Division Multiplexing (OFDM).

Multicarrier communication was first implemented in Frequency Division Multiplexing (FDM) in the early 1900's. In FDM, multiple low rate signals were transmitted using separate carrier frequencies for each signal. The various carrier frequencies had to be spaced sufficiently apart to avoid overlap of spectra and to be able to be efficiently separated at the receiver by using low cost filters. The empty spectral regions between the carrier frequencies led to very low spectral efficiency, but by breaking up the wide-band channel into several parallel narrower sub-channels, the effect of inter-symbol-interference (ISI) caused due to the frequency selective nature of the channel is greatly mitigated compared to the single channel wideband communication scheme. In time domain, the same can be explained as a method of achieving high immunity against multipath dispersion since the symbol duration on each sub-channel will be much larger than the channel time dispersion. Hence, the effects of ISI will be minimized. This gets rid of the need for expensive and complex equalization techniques. Also, due to the much narrower bandwidth of each sub-channel, effects of impulsive noise are also reduced. But to implement FDM, which yields the above mentioned benefits, a dedicated set of filters and oscillators are needed for each sub-channel, which makes the system expensive and complex to implement.

The Kineplex system developed by Collins Radio Co. [4] was one of the first algorithms to address the bandwidth efficiency problem of multicarrier transmission for data transmission over a high frequency radio channel subject to severe multi-path fading. Twenty tones spaced at frequency intervals almost equal to the signalling rate were used. The tones are selected in such a way that they can be separated at the receiver. A subsequent multi-tone system [5] was proposed using 9-point QAM constellations on each carrier, with correlation detection employed at the receiver.

The above techniques provide the orthogonality needed to separate multi-tone signals, but due to the infinite range of the spectrum of each component, the aggregate overlap of a large number of sub-channel spectra is pronounced. Also, spectrum spillage outside the allotted bandwidth is significant. With this in mind, it is desirable for each of the signal components to be bandlimited. There will still be overlap but with only the immediately adjacent sub-carriers, while still remaining orthogonal to them.

The first OFDM scheme was proposed by Chang in 1966 [6] for dispersive fading channels. Since then OFDM systems have been extensively employed [7, 8, 9, 10]. Saltzberg [11] studied a multi-carrier system employing orthogonal time-staggered QAM for the carriers. Use of DFT to replace the banks of sinusoidal generators and demodulators was suggested by Weinstein and Ebert [12] in 1971, which significantly reduced the implementation complexity of OFDM systems. In 1980, Hirosaki [13] introduced the DFT-based implementation of Saltzberg's O-QAM OFDM system.

The simplicity of OFDM has been recognized as an advantage to aid in its implementation [14, 15, 16]. The incoming data stream is converted from serial to $N$ parallel data streams and each parallel data stream is then modulated onto separate carriers using fast Fourier transforms (FFT), ensuring orthogonality. Due to the advancement in digital circuitry, the hardware to implement FFT is fast and inexpensive, making this scheme very attractive. Further, by using $N$ parallel data streams modulated by separate carriers instead of a single high rate stream modulated by a single carrier, the wide bandwidth of the channel is now broken down into $N$ narrow bandwidth channels which only experience flat fading. This avoids the need for equalizers at the receiver even over dispersive channels. To summarize, OFDM provides the following advantages over traditional FDM methods:

- High spectral efficiency due to the absence of guard bands

- Simple and efficient hardware realization by implementing the FFT operation

- Avoids inter-symbol-interference and thereby leads to low complexity receivers due to the avoidance of equalizers

- Each sub-carrier can have a different modulation/coding scheme leading to the design of highly robust adaptive transmission schemes

- Enables frequency diversity by spreading the subcarriers across the usable spectrum

- Provides good resistance against co-channel interference and impulsive noise

   Though OFDM offers the above advantages, it has some disadvantages:

- High sensitivity to Doppler shifts, requiring accurate frequency and time synchronization

- High Peak-to-Average Power Ratio due to the overlap of a large number of modulated sub-carrier signals which requires the transmit power amplifier to be linear across the whole signal range, or otherwise leads to clipping of peaks causing distortions. If the transmit power amplifier is not linear across the whole range, the out of band power leakage is significant which causes inter-carrier interference

- Loss in spectral efficiency due to the use of guard interval/cyclic prefix

With the substantial advancements in digital signal processing technology and drop in hardware costs, the presence of OFDM in telecommunication standards is rapidly growing. OFDM is used in broadcast standards such as Digital Video Broadcasting — Terrestrial (DVB-T) for international television with 1705 or 6817 subcarrier OFDM, Digital Multimedia Broadcasting (DMB) for use in multimedia data transfer for mobile devices in Korea, and Integrated Services Digital Broadcasting (ISDB) for digital television in Japan with Band Segmented Transmission (BST)-OFDM. Wireless network standards such as IEEE 802.11a, wireless local area networks (WLAN), metropolitan area networks (MAN), wireless personal area networks (WPAN) and HiperLAN/2 are based on OFDM transmissions. The IEEE P1901 draft standard for broadband over power line networks includes OFDM in its specifications.

The rest of the book is organized as follows. In Chapter 2, wireless communication channels are first introduced. Following this a baseband OFDM system is defined in Chapter 3. In Chapter 4 and Chapter 5, the two main pathologies of OFDM communication, namely carrier frequency offset (CFO) and high peak to average power ratio (PAPR) are presented along with techniques to mitigate their effects on the performance of OFDM systems. In Chapter 6, we provide code to simulate the error rate performance of a simple OFDM system. Following this, we also illustrate the effects of CFO and PAPR on the error rate performance of an OFDM system. All programs are written in MATLAB®.

CHAPTER 2

# Modeling Wireless Channels

In this chapter, some of the basic characteristics of the wireless channel are reviewed. In the first part of the chapter, characteristics of frequency-flat fading channels are introduced. In the second part of the chapter, frequency selective channels are introduced. A major reason to use OFDM is to mitigate frequency selective channels effectively. As will be shown in subsequent channels, OFDM converts one frequency selective channel into several frequency-flat fading channels, motivating the need for understanding the nature of frequency-flat fading channels, which is addressed in Section 2.1 and Section 2.2. The effect of frequency selectivity is addressed in Section 2.3, since OFDM is a frequency domain modulation scheme.

## 2.1 BASIC CHARACTERISTICS OF MOBILE RADIO CHANNELS

In mobile radio communication, the emitted electromagnetic waves may not reach the receiving antenna directly due to the obstacles blocking the line-of-sight path. The received waves are a superposition of waves coming from different directions due to reflection, diffraction, and scattering caused by buildings, trees, and other obstacles. This effect is known as multipath propagation.

In mobile communication the signal power drops off at the receiver due to (i) mean path loss, (ii) macroscopic fading, also called shadowing, and (iii) microscopic fading, also referred to as small scale fading. *Mean path loss* arises from inverse square law of power loss and depends on the distance of the traveling wave. *Macroscopic fading* or *shadowing* results from a blocking effect by obstacles such as buildings, large trees and mountains. *Microscopic* or *small scale* fading arises due to the *multipath propagation* where the received signal consists of an infinite sum of attenuated, delayed and phase-shifted replicas, caused due to the scattering of the transmitted signal by obstructions. Multipath propagation and the mobility of the receiver result in the spreading of the signal in different dimensions. These are mainly *delay spread* due to the presence of *resolvable* multipath components in time and *Doppler spread* in frequency due to the mobility of the terminal. We now describe the statistics of small scale fading along with the time and frequency spread that the channel introduces.

## 2.2 MICROSCOPIC OR SMALL SCALE FADING

Small scale fading refers to the rapid fluctuations of the received signal in space, time and frequency [17]. Since fading is caused by the superposition of a large number of independent scattered components, the in-phase and quadrature components of the received signal can be assumed to be independent zero mean Gaussian processes. Therefore, if no line-of-sight (LOS) path exists, the received signal consists only of sum the independent scattered components. The envelope, $|h|$, of the received signal has a Rayleigh density function given by

$$f_{|h|}(u) = \frac{2u}{\sigma_h^2} \exp\left(\frac{u^2}{\sigma_h^2}\right), \quad u \geq 0, \tag{2.1}$$

where $\sigma_h^2 := E[|h|^2]$. If there exists a line-of-sight (LOS) path between the transmitter and the receiver, the signal envelope is no longer Rayleigh distributed, but has a Ricean distribution. The Ricean distribution is defined in terms of the Ricean factor, $K$, which is the ratio of the power in the mean component of the channel to the power in the scattered (diffused) component. The Ricean probability distribution function (PDF) of the envelope of the received signal is given by

$$f_{|h|}(u) = \frac{2u}{\sigma_h^2} \exp\left(\frac{-(u^2 + \sigma_0^2)}{\sigma_h^2}\right) I_0\left(\frac{2u\sigma_0}{\sigma^2}\right), \quad u \geq 0, \tag{2.2}$$

where $\sigma_h^2 = E[|h - \sigma_0|^2]$ is the average power of non-line-of-sight component and $\sigma_0^2$ is the average power of the LOS component, the Ricean factor $K = \sigma_0^2/\sigma_h^2$, and $I_0$ is the modified Bessel function of the first kind defined as

$$I_0(x) = \frac{1}{\pi} \int_0^\pi \exp\left(-x\cos\theta\right) d\theta. \tag{2.3}$$

In the absence of a direct path, i.e., with $K = 0$, the Ricean PDF in (2.2) reduces to the Rayleigh PDF in (2.1) with $I_0(0) = 1$. There are other fading models, such as Nakagami fading or Weibull fading [18], which will not be considered.

An example of a multipath channel is shown in Figure 2.1. Out of several possible paths emanating from the transmitter, four are shown. There is one LOS path directly from the transmitter to the receiver. Three other paths shown from the transmitter, first encounter obstacles. Two of them reflect and reach the receiver, while the third reflects off an obstacle, but away from the receiver. These four paths are examples of actual paths. The received signal will consist of many such paths combining non-coherently at the receiver.

## 2.2.1 DOPPLER SPREAD: TIME SELECTIVE FADING

Due to relative motion between the transmitter and the receiver, the Doppler effect causes an apparent frequency shift of the received electromagnetic waves. If the angle of arrival of the $n$-th incident wave is $\theta_n$, the Doppler frequency shift of this component is given by $f_n := f_{\max} \cos\theta_n$, where $f_{\max} = (v/c)f_0$ is the maximum Doppler frequency, speed of the mobile unit is $v$, $c$ is the speed of light and the carrier frequency is $f_0$. Due to the Doppler effect, the spectrum of the transmitted signal undergoes a frequency expansion known as *frequency dispersion*. In time domain, the Doppler effect implies that the impulse response of the channel becomes time-variant. The *scattering function*, $S(\tau, f)$, can be used to capture the time-variant nature of the channel caused

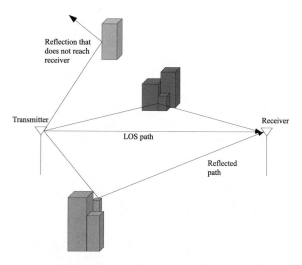

**Figure 2.1:** Illustration of a multipath channel. Time-delayed reflections of the same signal combine at the receiver.

by the Doppler effect [19]. The scattering function shows the Doppler power spectrum for paths with different delays $\tau$ and Doppler frequency $f$, and it is a complete characterization of the second order statistics of wireless channels [20]. Figure 2.2 illustrates a scattering function with respect to Doppler frequency $f$ and delay $\tau$. When averaged over the delay, $\tau$, the scattering function yields the Doppler spectrum, $S(f)$, which is the average power of the channel output as a function of the Doppler frequency:

$$S(f) = \int_{-\infty}^{\infty} S(\tau, f)d\tau. \tag{2.4}$$

The root mean square (RMS) bandwidth of $S(f)$ is called the *Doppler spread*, $f_{\mathrm{rms}}$, and is given by

$$f_{\mathrm{rms}} = \sqrt{\frac{\int_{\mathcal{R}_f} (f - f_{\mathrm{avg}})^2 S(f)df}{\int_{\mathcal{R}_f} S(f)df}}, \tag{2.5}$$

where $\mathcal{R}_f$ is the region where $f_0 - f_{\mathrm{max}} \leq f \leq f_0 + f_{\mathrm{max}}$ and $f_{\mathrm{avg}}$ is the average frequency of the Doppler spectrum given by

$$f_{\mathrm{avg}} = \frac{\int_{\mathcal{R}_f} f S(f)df}{\int_{\mathcal{R}_f} S(f)df}. \tag{2.6}$$

In the presence of direct path, the Doppler spectrum, $S(f)$, is modified by an additional discrete frequency component corresponding to the relative velocity between the base-station and the terminal. Fading introduced by the Doppler effect can be characterized by the coherence time, $T_c$, of the channel and is typically defined as the time lag at which the signal autocorrelation coefficient

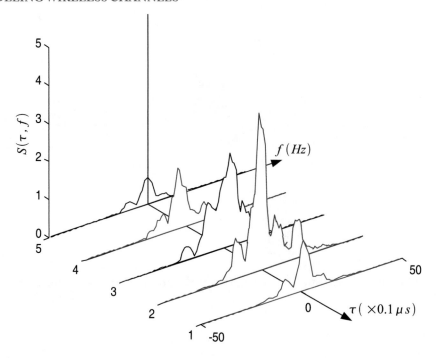

**Figure 2.2:** Plot of the scattering function, $S(\tau, f)$.

reduces to 0.7. The coherence time can also be approximated as the reciprocal of the Doppler spread, i.e., $T_c \approx 1/f_{\text{rms}}$. Thus, the coherence time serves as a measure of how fast the channel changes in time, i.e., the larger the coherence time, the slower the channel fluctuation.

The coherence time and the Doppler effect play an important role in the functioning of multicarrier systems. In a multicarrier system, a frequency selective channel with large bandwidth is divided into several narrow-band subcarriers. If the number of subcarriers increases for a given bandwidth, the bandwidth assigned to each channel reduces. This implies that the pulse width of the symbols in time increases. Therefore, the system has to designed carefully for the symbol pulse width to not exceed the coherence time of the channel. Doppler also causes loss of orthogonality of the subcarriers in frequency which leads to inter-carrier interference, and this will be covered in Chapter 4.

## 2.2.2   DELAY SPREAD: FREQUENCY SELECTIVE FADING

In multipath propagation, depending on the incident phase of the waves from each of the multiple paths, their superposition can be constructive or destructive. Moreover, there may exist multiple resolvable components depending on the transmission rate. Thus, the presence of more than one resolvable multipath component causes time dispersion of the transmitted pulse and often several individually distinguishable pulses occur at the receiver. This time, dispersion of the pulses manifests as frequency distortion in the frequency domain due to the non-flat frequency response of the channel. The distortion caused by multipath propagation is usually modeled as linear and often compensated by an equalizer in single carrier communication. In multicarrier communications, however, several narrow band parallel subcarriers are transmitted where each subcarrier is designed to observe frequency-flat fading.

The delay separation between paths increases with path delay [21]. The span of path delays between the first and the last replicas of the received signal is called the *delay spread*. The RMS delay spread of the channel, $\tau_{\text{rms}}$, is defined as

$$\tau_{\text{rms}} = \sqrt{\frac{\int_0^{\tau_{\max}} (\tau - \tau_{\text{avg}})^2 A(\tau) d\tau}{\int_0^{\tau_{\max}} A(\tau) d\tau}}, \tag{2.7}$$

where the *multipath intensity profile* or *power delay profile*, $A(\tau)$, is the average power of the channel output as a function of delay $\tau$, $\tau_{\max}$ is the maximum path delay and $\tau_{\text{avg}}$ is the average delay spread given by

$$\tau_{\text{avg}} = \frac{\int_0^{\tau_{\max}} \tau A(\tau) d\tau}{\int_0^{\tau_{\max}} A(\tau) d\tau}. \tag{2.8}$$

The multipath intensity profile is related to the spectrum $S(f)$ as

$$A(\tau) = \int_{-\infty}^{\infty} S(\tau, f) df. \tag{2.9}$$

Therefore, to avoid inter-symbol interference (ISI) in linearly modulated systems, the symbol duration, $T \gg \tau_{\text{rms}}$ should be satisfied. In the OFDM scenario, the symbols being transmitted are separated by a specialized guard band called the cyclic prefix. The length of the cyclic prefix should be at least as long as the maximum delay spread. The cyclic prefix, and its role in OFDM systems, is explained in more detail in Chapter 3. In the presence of delay spread, the channel can be modeled as a tapped delay line filter and, consequently, frequency-selective fading is experienced. Frequency-selective fading can be characterized in terms of its *coherence bandwidth*, $B_c$, which is the frequency difference for which the channel's autocorrelation coefficient reduces to a prescribed value (example, 0.7 in [22]). The coherence bandwidth is a measure of the channel's frequency selectivity and is the reciprocal of the RMS delay spread, i.e., $B_c \approx 1/\tau_{\text{rms}}$. The power delay profile is often modeled as one-side exponential distribution:

$$A(\tau) = \frac{1}{\tau_{\text{avg}}} \exp\left(-\tau/\tau_{\text{avg}}\right), \quad \tau \geq 0. \tag{2.10}$$

Using (2.7), it can be shown that for the exponential delay profile given in (2.10), $\tau_{\text{rms}} = \tau_{\text{avg}}$. Typically, delay spread, $\tau_{\text{rms}}$, increases with distance from the terminal. This is due to the fact that at larger distances, multipaths with large delays have strengths comparable to the direct path

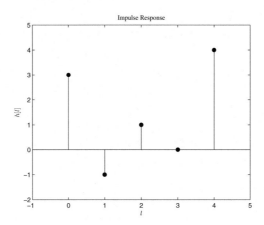

**Figure 2.3:** Impulse response of channel in (2.12).

which ultimately increases $\tau_{\text{rms}}$. In flat rural areas, $\tau_{\text{rms}}$ is less than 0.05 $\mu s$, in urban areas $\tau_{\text{rms}}$ is approximately $0.2\mu s$ and in hilly terrains $\tau_{\text{rms}}$ is around 2-3 $\mu s$ [23]. In a multicarrier system, a frequency selective channel is divided into several narrow-band subcarriers. The subcarriers are chosen such that each of them is a frequency-flat fading channel. The values of the RMS delay spread and the coherence bandwidth play an important role in determining the number of subcarriers to be used. For instance, consider a system with a total bandwidth of $BW = 2\text{MHz}$. The system is deployed in an environment that has an RMS delay spread of $25\mu$s or a coherence bandwidth of $B_c = 40\text{kHz}$. For channels to be frequency-flat fading, the required coherence bandwidth on each of $N$ subcarriers is given by $B_N := BW/N \ll B_c$. If $B_N = 0.1B_c$, at least $N = 500$ subcarriers have to be used. In OFDM, as will be shown later, it is preferred that the number of subcarriers be a power of 2, in which case, $N = 512$ can be used.

## 2.3   TAPPED DELAY LINE MODEL FOR FREQUENCY SELECTIVE FADING CHANNELS

In this section, we consider frequency selective channels and briefly discuss some methods to mitigate the effects of frequency selective channels. The drawbacks of these schemes are presented in order to motivate the need for multicarrier systems such as OFDM.

Frequency selective channels are commonly represented using the tapped delay line model. In a tapped delay line model, a data line is tapped at different time delays, weighted with different values, and then summed together to provide an output. Such a model efficiently represents data received via multiple paths for a signal from the same source, making it a good fit for frequency selective channels. For a frequency selective channel represented using $L$ taps, if the transmitted

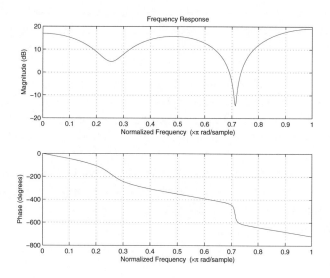

**Figure 2.4:**  Frequency response of channel in (2.12).

data is $u[n]$, the output at the receiver, $r[n]$, is represented as [20]

$$r[n] = \sum_{l=0}^{L-1} h[n; l]u[n - l],$$ (2.11)

where $h[n; l]$, $l = 0, \ldots, L - 1$ represent the $L$ taps of the frequency selective channel at time $n$. This convolutional channel can also be interpreted as an FIR filter of order $L - 1$. For a frequency selective fading channel, the channel coefficients are modeled as random [1].

As an example, consider a channel $h[n; l]$ at a fixed instant of time. Assuming that the channel is time invariant, we drop the time index. Consider, as an example, the channel whose impulse response can be represented as

$$h[l] = 3\delta[l] - \delta[l - 1] + \delta[l - 2] + 4\delta[l - 4].$$ (2.12)

For this channel, the impulse response and frequency response are shown in Figure 2.3 and Figure 2.4, respectively. We can see from Figure 2.3 that the channel represents a multipath channel, and from Figure 2.4, we can see that the response of the channel is not the same at each frequency, making it a frequency selective channel. In multicarrier systems such as FDM and OFDM, the frequency spectrum is divided into several narrow-band channels called subcarriers. If the channel bandwidths are small, each can be considered to be a frequency-flat fading channel. While this is a

---

[1]It should be noted that in some cases, especially in wired ISI channels, such as telephone lines, the channel taps are modeled as deterministic [24].

good approach to mitigate the effects of a frequency selective channel, a subcarrier that occurs at a trough on the frequency response of the channel will result in a channel with very poor performance. Strategies such as error-control coding across subcarriers [25] are used to improve performance in such situations.

As shown in (2.11), transmission over a frequency selective channel can be considered as a convolution in time between the data and the tapped delay line representation of the channel. The Viterbi algorithm considers the channel as a state-machine and can be used to decode the data, and it is shown to provide the maximum-likelihood solution [24]. However, the Viterbi algorithm grows exponentially in the number of channel taps.

Alternatively at the receiver, the convolution in (2.11) can be inverted in order to estimate the transmitted data, in a process called equalization. Several suboptimal techniques can be used for equalization. Linear equalization uses an FIR filter, $g[l]$, to estimate the value of the transmitted symbol, $u[n]$, to yield the estimate:

$$\widehat{u}[n] = y[n] * g[n]. \tag{2.13}$$

The filter, $g[l]$, has to be selected so that the estimate, $\widehat{u}[n]$, is close to the transmitted signal, $u[n]$. In the absence of channel noise, $g[l]$ is selected such that

$$h[l] * g[l] = \delta[l], \tag{2.14}$$

so that $\widehat{u}[n] = u[n]$. Since the convolution of two FIR filters will never yield $\delta[l]$, selecting $g[l]$ to satisfy (2.14) is not possible with an FIR equalizer. Instead, the optimum coefficients of $g[l]$ are chosen in a way to minimize a performance index, such as the mean square error (MSE) between the transmitted symbol, $u[n]$, and the estimate of the symbol, $\widehat{u}[n]$, at the receiver as follows:

$$g^{opt}[l] = \operatorname*{argmin}_{\{g[l]\}} E\left[|u[n] - \widehat{u}[n]|^2\right]. \tag{2.15}$$

More information about these and other more complex equalizers such as the decision-feedback equalizer (DFE) and iterative solutions to (2.15) can be found in [24].

Mitigation of the effects of the frequency selective channel requires estimation of the taps for both equalization and the Viterbi algorithms. Due to the convolutional nature of the channel, channel estimation cannot be performed by transmitting a pilot tone. A white noise sequence is transmitted and cross-correlated with the received signal in order to estimate the channel [26]. In contrast, with frequency-domain schemes such as FDM or OFDM, the frequency-flat fading channel on each subcarrier can be estimated individually by transmitting a pilot tone at each subcarrier. If the entire channel estimate is required with a few pilots, interpolation of the channel estimates in the frequency domain will yield the required result. The structure of an OFDM system, which allows such estimation, is discussed in Chapter 3.

CHAPTER 3

# Baseband OFDM System

In this chapter, the basic model of the OFDM system is introduced. First, an analog interpretation of the OFDM system is presented. Following this, the discrete symbol-rate sampled OFDM transmission scheme is developed. Block transmissions built on a matrix-vector framework are introduced, which subsume transmission schemes that use zero padding (ZP), OFDM using cyclic prefix, and pre-coded transmissions. We will also discuss OFDM as a block transmission scheme effective in mitigating ISI in large delay spread environments.

## 3.1    INTRODUCTION TO OFDM

As discussed in Chapter 2, a frequency selective channel has a convolutional effect on transmitted data, and methods such as the Viterbi algorithm or equalization are used to mitigate the effects of the frequency selective channel. Orthogonal Frequency Division Multiplexing (OFDM) is a technique that can also be used to mitigate frequency selective channels.

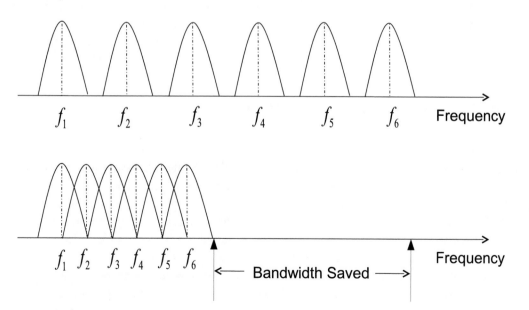

**Figure 3.1:** An example of FDM transmissions.

In a simple frequency division multiplexing (FDM) system, the entire channel bandwidth is divided into several narrow bandwidth channels, referred to as subcarriers. If the bandwidth of the subcarrier is suitably small, it can be considered to be a flat fading channel. In an FDM system, the subcarriers need to be assigned in such a way that they do not interfere with each other. Such a system is shown in the top half of Figure 3.1 where the allotted bandwidth is partitioned into subcarriers. To make allowances for bandwidths that are not restricted in frequency, and for filters, the subcarriers are spaced sufficiently apart from each other. The restriction stops us from utilizing a partitioning system as shown in the bottom half of Figure 3.1.

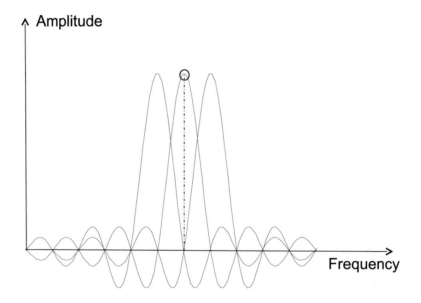

**Figure 3.2:** OFDM symbols represented using sinc functions.

In contrast, in an OFDM system, in addition to dividing the frequency spectrum into separate parts, they are shaped as well, as shown in Figure 3.2. Due to this shaping, when a subcarrier is sampled at its peak, all other subcarriers have zero-crossings at that point, and they do not interfere with the subcarrier being sampled. In case this sampling is off-peak, there could be interference from adjacent subcarriers. Furthermore, not truncating the spectrum of each subcarrier reduces the demands on filters, and it allows the symbols to be restricted in time. In a typical OFDM system, data symbols are transmitted over each subcarrier and received without interference.

To implement such a system, the symbols are first considered in frequency. By taking the IFFT of the data symbols, time-domain representations are obtained. A cyclic prefix is added to this representation in time. An interval of the time-representation of the symbols is copied and added to the front, comprising the cyclic prefix. This data, after the addition of the cyclic prefix, is transmitted over the frequency selective channel. At the receiver, the cyclic prefix is dropped, and the FFT of the rest provides the symbols at the receiver [27].

In this process, the length of the cyclic process plays an important role. The length of the cyclic prefix is chosen such that it is larger than the maximum delay spread of the wireless channel. Figure 3.3 helps understand the significance of the cyclic prefix. It shows three subcarriers in the

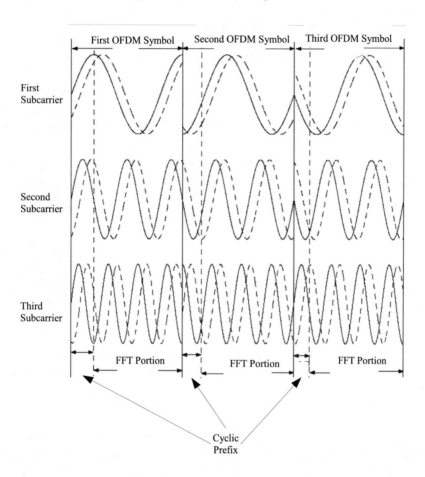

**Figure 3.3:** Importance of Cyclic Prefix.

time domain after passing through a two-ray channel environment ($L = 2$ in (2.11)). The solid curves represent the subcarriers that have reached the receiver without any delay, and the dotted ones represent those that have reached after a certain delay. Of course, what we see at the receiver is a sum of the signals. Figure 3.3 also shows the phase transitions that might occur at symbol interval boundaries. Since the choice of the cyclic prefix interval is larger than the delay spread, the delayed replicas of the subcarriers show phase transitions within the guard interval. At the receiver, since FFT is taken after discarding the guard interval part of the received signal, the orthogonality between any subcarrier and delayed version of any other subcarrier is still preserved [28].

The analog method described provides good intuition into the working of an OFDM system. However, in the case of digital systems, the continuous-time methods described cannot be used. Digitization and the use of block transmissions are required. These digitization techniques are used to formally introduce the concepts of OFDM later on in the chapter. Additionally, the discrete model is more suited for simulation using computer programs.

## 3.2   DISCRETE BASEBAND BLOCK TRANSMISSIONS

The purpose of this section is to establish a convenient discrete-time framework encompassing well-known block transmission techniques like OFDM with a Cyclic Prefix (CP-OFDM), zero-padded (ZP) transmissions, and block pre-coded transmissions that process information symbols in blocks. We will also show that block-transmissions are an effective way to mitigate channel induced ISI [29]. This unifying model is useful in holding a signals-and-systems view of the entire transmission process and is also used in describing the OFDM transmission technique later in this chapter.

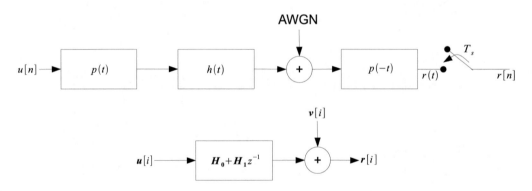

**Figure 3.4:** Serial Transmissions (above) and Block Transmissions (below).

We first begin with a linearly modulated transmission system over a frequency selective channel. In Figure 3.4, $u[n]$, $n \in \mathbb{Z}$ are pulsed shaped by a filter with response $p(t)$ and then sent over a wireless channel with an impulse response $h(t)$, and additive noise, $v(t)$. The received signal is then passed through a filter with response $p(-t)$, matched to the transmit pulse-shaping filter $p(t)$. The equivalent received discrete-time sequence is given by

$$r[n] = \sum_{l=0}^{L-1} u[n-l]h[l] + v[n], \tag{3.1}$$

where $h[n] := h(nT_s)$, $r[n] := r(nT_s)$, $v[n] := v(nT_s)$, $T_s$ is the sampling period, and $h(t)$, $r(t)$ and $v(t)$ are the analog-time representations of the channel, received symbols, and additive channel noise, respectively. Let us now link this serial transmission setup with a block transmission setup. In block transmissions, blocks of length $P$ are obtained from the symbols $u[n]$ such that $P \gg L$. Let $\mathbf{u}[i]$ denote the $i^{th}$ transmitted block[1] which is equal to $[u[(i-1)P], u[(i-1)P+1], \ldots, u[(i-1)P + P - 1]]^T$. Using (3.1), it can be shown that

$$\mathbf{r}[i] = \mathbf{H}_0\mathbf{u}[i] + \mathbf{H}_1\mathbf{u}[i-1] + \mathbf{v}[i], \qquad (3.2)$$

where $\mathbf{r}[i] = [r[(i-1)P], \ldots, r[(i-1)P+P-1]]^T$ and $\mathbf{v}[i] = [v[(i-1)P], \ldots, v[(i-1)P+P-1]]^T$ are the received and noise vectors, respectively, in the $i^{th}$ block interval and because $P \gg L$, the $P \times P$ channel matrices $\mathbf{H}_0$ and $\mathbf{H}_1$ are given by

$$\mathbf{H}_0 = \begin{pmatrix} h[0] & 0 & 0 & \ldots & 0 \\ \vdots & h[0] & 0 & \ldots & 0 \\ h[L-1] & \ldots & \ddots & \ldots & 0 \\ \vdots & \ddots & \ldots & \ddots & 0 \\ 0 & \ldots & h[L-1] & \ddots & h[0] \end{pmatrix}_{P \times P}, \qquad (3.3)$$

and

$$\mathbf{H}_1 = \begin{pmatrix} 0 & \ldots & h[L-1] & \ldots & h[1] \\ \vdots & \ddots & 0 & \ddots & \vdots \\ 0 & \ldots & \ddots & \ldots & h[L-1] \\ \vdots & \vdots & \vdots & \ddots & \vdots \\ 0 & \ldots & 0 & \vdots & 0 \end{pmatrix}_{P \times P}, \qquad (3.4)$$

and shown in the lower block diagram in Figure 3.4. It may be noted that similar to ISI in the serialized transmission shown in (3.1), there is *inter-block interference* (IBI) for the block $\mathbf{u}[i]$, but only from the immediately preceding block $\mathbf{u}[i-1]$, due to causality. It is easy to see that this is a consequence of choosing $P \gg L$. The matrix-vector framework given in (3.2) unifies many well-known transmission schemes in the following way:

- **Block-Precoded Transmissions** – Linear precoded transmissions, such as those that have been proposed for OFDM in [30] are instances of block transmission. They involve linearly coding the information block $\mathbf{s}[i]$ by multiplying with a precoding matrix $\mathbf{Th}$. With linear precoding, $\mathbf{u}[i]$ is derived from $\mathbf{Th} \cdot \mathbf{s}[i]$ after either zero-padding or after taking FFT and appending a cyclic prefix and is transmitted over the channel, which will be discussed next.

- **Zero-Padded Transmissions (ZP)** – Let us recollect from (3.2) that there is IBI in the received block at the $i^{th}$ time instant because of the last $L - 1$ symbols in $\mathbf{u}[i-1]$. It is easy

---

[1]The term "symbol" is used interchangeably with "block" in block transmissions.

to see that this IBI can be made equal to 0 by making the last $L - 1$ or more of the symbols in $\mathbf{u}[i - 1]$ equal to 0. This ZP transmission scheme [31] can be mathematically represented as

$$\mathbf{y}[i] = \mathbf{H}_0 \mathbf{T}_{\text{zp}} \mathbf{u}[i] + \mathbf{v}[i], \tag{3.5}$$

where $\mathbf{u}[i] = \mathbf{T}_{\text{zp}} \mathbf{s}[i]$ is obtained from an $N \times 1$ block of symbols by appending $\bar{L} > L$ zeros through a matrix operator $\mathbf{T}_{\text{zp}} := [\mathbf{I}_N^T \ \mathbf{0}_{\bar{L} \times N}^T]^T$ which appends $\bar{L}$ zeros to $\mathbf{s}[i]$. This period of "silence" at the end of the $i^{th}$ block prevents IBI in the $(i + 1)^{th}$ block, since $\mathbf{H}_1 \mathbf{T}_{\text{zp}} = \mathbf{0}$.

- **Cyclic-Prefixed OFDM (CP-OFDM)** – Cyclic-prefixed OFDM [29, 32, 33, 34], which is a very popular Multicarrier (MC) modulation scheme, is also a block transmission scheme and fits very well into the data model developed in (3.2). In OFDM, blocks of data, $\mathbf{s}[i]$, of length $N$, are obtained from a serial stream of input symbols, $s[n]$. In each block interval $i$, the $N$ elements in $\mathbf{s}[i]$ are modulated onto $N$ subcarriers. This is achieved through discrete IFFT at the transmitter. In order to prevent IBI, guard intervals between blocks of symbols are introduced. But instead of not transmitting anything in the guard interval duration like the zero-padded transmission scheme described in (3.5), the last $\bar{L}$ data-points in the tail portion of the OFDM symbol are transmitted and termed "cyclic-prefix." At the receiver, the cyclic prefix portion of the received signal is discarded, and FFT is taken on the remainder. At this stage, the relation between the input block $\mathbf{s}[i]$ and the output of the FFT appears as follows:

$$\mathbf{y}[i] = \mathbf{H}\mathbf{s}[i] + \mathbf{v}[i], \tag{3.6}$$

where $\mathbf{v}[i]$ is the noise vector. Addition of the cyclic prefix makes $\mathbf{R}_{\text{cp}} \mathbf{H}_0 \mathbf{T}_{\text{cp}}$ into a circulant matrix, which results in the channel matrix, $\mathbf{H} = \mathbf{F}_N \mathbf{R}_{\text{cp}} \mathbf{H}_0 \mathbf{T}_{\text{cp}} \mathbf{F}_N^H$ being diagonal. Therefore, there is no ISI between the elements of $\mathbf{y}[i]$ in (3.6). Later in the chapter, we shall see in detail all the transmitter and receiver operations that result in the data model in (3.6) for OFDM. We will also address the explicit derivation of (3.6) and the relationship between $h[l]$ and $\mathbf{H}$.

## 3.3 DISCRETE-TIME OFDM MODEL

Figure 3.5 shows a block-diagram representation of the discrete-time implementation of OFDM. The transmissions occur on a wireless ISI channel, $h(t)$, which is modeled as a tap-delay-line filter with $L$ taps, a maximum delay spread of $\tau_{\text{max}}$, average delay spread of $\tau_{\text{avg}}$, and RMS delay spread of $\tau_{\text{rms}}$. All the transmitter operations like serial-to-parallel conversion of the input data, taking IFFT on it, cyclic-prefix insertion, pulse-shaping and transmitting on the channel shown in Figure 3.5 are for the $i^{th}$ block interval and are explained in detail below:

- **Blocking** – It has been shown in Section 3.2 that OFDM fits into the general class of block transmissions. The OFDM modulator parses a continuous stream of input data into blocks of length $N$, as shown in Figure 3.5. Later, as we shall see, this blocking of input data and further processing helps in countering the channel induced ISI. Figure 3.5 shows the signal

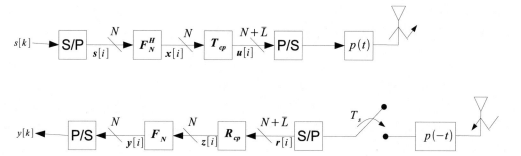

**Figure 3.5:** Discrete-Time Implementation of OFDM.

processing that takes place in the $i^{th}$ OFDM symbol interval during which the $N$ uncoded data elements, $s[k]$, $k = (i - 1)N, (i - 1)N + 1, \ldots, (i - 1)N + N - 1$, are grouped into an OFDM symbol, $\mathbf{s}[i]$, of length $N$. The OFDM symbol $\mathbf{s}[i]$ is then subjected to further processing.

- **Subcarrier Modulation** – In its discrete-time implementation, the modulation of subcarriers by the data is achieved through an IFFT operation. In Figure 3.5, we see that the $N$ data elements, $\mathbf{s}[i]$ are subjected to an IFFT operation.

- **Cyclic-prefix insertion** – An important operation that helps in preserving the orthogonality of the subcarriers is the insertion of cyclic-prefix between OFDM symbols. The number of symbols in the cyclic prefix is at least as many as the number of taps in the FIR filter representation of the frequency selective channel.

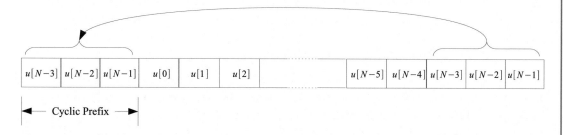

**Figure 3.6:** Cyclic prefix added to a block of data.

The output, after taking IFFT and inserting cyclic-prefix, is $\mathbf{u}[i] = [u[(i - 1)P], u[(i - 1)P + 1], \ldots, u[(i - 1)P + P - 1]]^T$, where

$$u[n] = \frac{1}{\sqrt{N}} \sum_{k=(i-1)N}^{(i-1)N+N-1} s[k] \exp\left(j2\pi nk/N\right), \qquad (3.7)$$

for $n = (i-1)P, (i-1)P + 1, \ldots, (i-1)P + P - 1$ and $P = N + \bar{L}$. Alternatively, the generation of $\mathbf{u}[i]$ from $\mathbf{s}[i]$ can also be described as

$$\mathbf{u}[i] = \mathbf{T}_{\text{cp}}\mathbf{F}_N^H\mathbf{s}[i], \qquad (3.8)$$

where $\mathbf{T}_{\text{cp}} := \begin{bmatrix} \mathbf{I}_{\text{cp}}^T & \mathbf{I}_N^T \end{bmatrix}^T$ is a cyclic prefix inserting matrix with $\mathbf{I}_{\text{cp}}$ being the last $\bar{L}$ rows of the $N \times N$ identity matrix $\mathbf{I}_N$, $\mathbf{F}_N$ is an $N \times N$ DFT matrix, and $\mathbf{F}_N^H$ is an $N \times N$ IDFT matrix obtained by taking the Hermitian of $\mathbf{F}_N$. An OFDM symbol with the CP added to it is shown in Figure 3.6.

- **Pulse-Shaping** – Samples $u[n], n = (i-1)P, (i-1)P + 1, \ldots, (i-1)P + P - 1$ of the $P \times 1$ OFDM symbol $\mathbf{u}[i]$ (also containing the cyclic-prefix) that we see in Figure 3.5 are pulse shaped with a transmit filter, $p(t)$, and transmitted over the channel, $h(t)$. Practical issues like out-of-band energy emissions, inter-channel interference and peak-to-average power dictate the choice of the exact pulse-shaping scheme to be used. OFDM systems can be categorized into two classes depending on the pulse shaping filter used: (i) the class of OFDM systems that use time-limited pulses [33, 35], typically rectangular pulses that overlap in the spectral domain, but are orthogonal and (ii) the class of OFDM systems designed with infinitely long pulses, but they are realized with their truncated versions. Orthogonality conditions for the second category were presented in [6] and an application with offset QAM was presented in [11].

Even though the data samples, $u[n]$, are those of an OFDM system, the channel, $h[n]$, affects them the same way it does other single-carrier transmission schemes. Therefore, the continuous-time signal that is received is the same as in (3.1). At this stage, the data model for the $i^{th}$ block at the receiver, $\mathbf{r}[i]$, is exactly the same as shown in (3.2). After this, the receiver simply eliminates the IBI due to $\mathbf{H}_1$ by discarding the first $\bar{L}$ samples received during the guard interval and performs an FFT on the remainder. All the aforementioned operations on $\mathbf{r}[i]$, can be mathematically described as below:

$$\mathbf{y}[i] = \mathbf{F}_N\mathbf{R}_{\text{cp}}\mathbf{r}[i] = \mathbf{F}_N\mathbf{R}_{\text{cp}}\mathbf{H}_0\mathbf{T}_{\text{cp}}\mathbf{F}_N^H\mathbf{s}[i] + \mathbf{F}_N\mathbf{R}_{\text{cp}}\mathbf{v}[i], \qquad (3.9)$$

where $\mathbf{R}_{\text{cp}} := [\mathbf{0}_{N \times \bar{L}}\mathbf{I}_N]$ removes the initial cyclic prefix part of the received symbol. Note that the IBI term because of $\mathbf{H}_1$ is also made zero since $\mathbf{R}_{\text{cp}}$ removes its first $\bar{L}$ rows. The advantage of employing IFFT and FFT at the transmitter and receiver, respectively, is that the factor $\mathbf{F}_N\mathbf{R}_{\text{cp}}\mathbf{H}_0\mathbf{T}_{\text{cp}}\mathbf{F}_N^H$ simplifies to a diagonal matrix $\mathbf{H}$ which can easily be inverted (provided the inverse exists) [29]. Since $\mathbf{H}$ is a diagonal matrix, the input-output relation at any subcarrier $k$ is a simple one without any ISI and is as follows:

$$y[k] = H[k]s[k] + v[k], \qquad (3.10)$$

for $k = (i-1)N, (i-1)N + 1, \ldots, (i-1)N + N - 1$, where

$$v[k] := N^{-1/2} \sum_{(i-1)N}^{(i-1)N+N-1} v[n] \exp\left(-j2\pi kn/N\right), \qquad (3.11)$$

and $H[k]$ representing the channel gain of the $k^{th}$ subcarrier is the $k^{th}$ element on the principal diagonal of $\mathbf{H}$ is given by

$$H[k] = \sum_{n=0}^{L-1} h[n] \exp\left(-j\frac{2\pi kn}{N}\right). \tag{3.12}$$

Looking at the input-output relation in (3.10), it is clear that through IFFT and cyclic-prefix insertion at the transmitter and with matching operations at the receiver, OFDM has turned an ISI channel requiring potentially complex equalization at the receiver into a set of flat-fading channels. This is the single-most important advantage of OFDM: robustness to large delay spread environments obviating the need for complex equalization at the receiver. One drawback of this method is that when the gain of a subcarrier is low, equalization amplifies the additive noise. This problem is exacerbated when a subcarrier lies on a channel null, and the data transmitted over that subcarrier is completely lost. To mitigate this problem, error-control coding is used to code symbols across subcarriers [25].

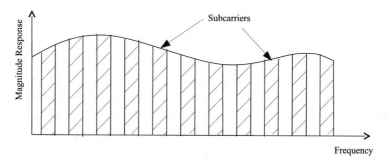

**Figure 3.7:** A frequency selective fading channel divided into orthogonal subcarriers. Alternate subcarriers are shaded for clarity.

This representation of OFDM as shown in (3.10) can be also interpreted as shown in Figure 3.7. It shows a frequency selective channel divided into subcarriers, with no overlap. Each subcarrier then behaves like a flat fading channel with no interference from other subcarriers. Data can be transmitted over each of these subcarriers independently. In this scenario, for channel equalization, it is only necessary to compensate for the effect of each subcarrier indivdually. In order to do this, the subcarriers gains have to be estimated. For OFDM, each subcarrier gain can be simply estimated with individual pilots since each subcarrier is now equivalent to a flat fading channel with no interference from other subcarriers [36]. In case the entire channel needs to be estimated, and not just the subcarriers, a simple interpolation will yield the required information.

OFDM is used in the DAB [1], the DVB-T [2], the DMB [3] and the IEEE 802.11a [37] standards. Typical values of parameters such as bandwidth, number of subcarriers, spacing of subcarriers, modulation schemes and bit rates used in these standards are shown in Table 3.1. Some of these standards work in multiple modes and bands. For instance, although the DAB standard can

| Table 3.1: Typical Values of OFDM parameters as used in Common Standards | | | | |
|---|---|---|---|---|
| **Standards** | | | | |
| **Name** | **DAB** | **DVB-T** | **DMB** | **IEEE 802.11a** |
| Bandwidth (MHz) | 174-240 <br> 1,452-1,492 | 174-240 <br> 470-862 | 470-862 | 4912-5825 |
| Number of Subcarriers | Mode I: 1,536 <br> Mode II: 384 <br> Mode III: 192 <br> Mode IV: 768 | 2K mode: 1705 <br> 8K mode: 6,817 | 3,7802 | 52 |
| Subcarrier Spacing (Hz) | Mode I: 1000 <br> Mode II: 4000 <br> Mode III: 8000 <br> Mode IV: 2000 | 2K mode: 4,464 <br> 8K mode: 1,116 | 2,000 | 312.5K |
| Modulation Scheme | $\pi/4$ DQPSK | QPSK | QAM | BPSK |
| Bit Rate (Mbits/s) | 0.576-1.152 | 24 | 4.81-32.49 | 6-54 |

operate above 30 MHz, it has spectra allocated for it in Band III (high-band VHF; 174–240 MHz) and L-band (1452–1492 MHz). DAB has a number of country specific transmission modes (I, II, III and IV). For worldwide operation, a receiver must support all 4 modes: (i) Mode I for Band III, Earth; (ii) Mode II for L-Band, Earth and satellite; (iii) Mode III for frequencies below 3 GHz, Earth and satellite; and (iv) Mode IV for L-Band, Earth and satellite. In the case of DVB-T, there are two choices for the number of carriers known as 2K-mode or 8K-mode. In the 2K mode, 1,705 (approximately 2000) subcarriers are used that are spaced approximately 4kHz apart. In the 8K mode, 6,817 carriers, approximately 1 kHz apart, are used. The DVB-T has been allocated frequencies in Band III and Band IV. For these standards that work in multiple modes and bands, receivers are generally marketed so that they can be set up to work with all the different systems.

CHAPTER 4

# Carrier Frequency Offset

Before an OFDM symbol can be successfully demodulated, the receiver has to synchronize to both the transmitted frame timing and carrier frequency. First, the receiver has to know where exactly it has to sample the incoming OFDM symbol prior to the FFT process. Secondly, the receiver has to estimate and correct for any carrier frequency offset because offset can result in inter-carrier-interference (ICI). In fact, the sensitivity to timing and carrier offset errors is higher in OFDM systems than in single carrier systems [34]. Transmitted signals are provided with timing, frequency, and phase reference parameters to assist with synchronization at the receiver. Proper detection at the receiver requires knowledge of these parameters. The first task of the receiver is to estimate symbol boundaries. If the receiver cannot clearly identify the symbol lengths, then ISI occurs. A preamble consisting of a sequence of known symbols is used for the receiver. Once the presence of symbol is detected, the next task is to estimate the frequency offset. Frequency offset occurs due to unmatched frequencies on the received signal and the local oscillator at the receiver. Therefore, subcarriers could be shifted from their original positions resulting in a non-orthogonal signal at the receiver resulting in ICI after the FFT due to the FFT output containing interfering energy from all other subcarriers. Other problems such as out-of-band radiation [38, 39] can also occur with OFDM transmissions.

## 4.1    CARRIER SYNCHRONIZATION ERROR

Frequency offsets are typically introduced by a (small) frequency mismatch in the local oscillators of the transmitter and the receiver. Doppler shifts can also induce a slight frequency change of the carrier frequency [40] and hence, lead to frequency mismatch.

The impact of a frequency error can be seen as an error where the received signal is sampled during demodulation. Figure 4.1 depicts this twofold effect.

Since the subcarriers (SC) are orthogonal, when viewed in time domain, the peak of any sinc is aligned with the zeros of all other sincs. Ideally, each SC is sampled at its peak, and there is no contribution from the other SCs. However, when there is a frequency offset, sampling may not occur at the peaks but at an offset point. The amplitude of the desired SC is reduced, and ICI arises from the adjacent SCs.

Here, we would like to recall that after parallel to serial conversion the output of the IFFT can be represented as

$$x[n] = \frac{1}{N} \sum_{k=0}^{N-1} s[k] \exp\left(\frac{j2\pi kn}{N}\right). \tag{4.1}$$

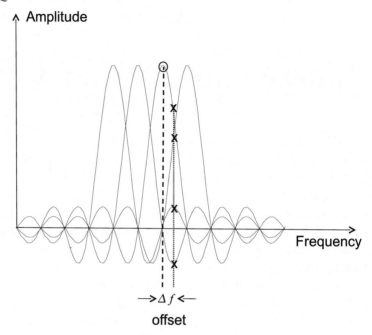

**Figure 4.1:** Sampling mismatch due to CFO.

We now consider the case where there exists a mismatch in the frequencies of the received signal and the local oscillator at the receiver. Ignoring the effects of the additive noise, the received signal after removal of CP can be written as

$$z[n] = \frac{1}{N} \sum_{k=0}^{N-1} s[k]H[k] \exp\left(\frac{j2\pi n(k + \Delta f)}{N}\right), \tag{4.2}$$

where $\Delta f$ represents the relative frequency offset defined as the ratio of the actual frequency offset to the intercarrier spacing, and $H[k]$ is the transfer function of the channel at the frequency of the $k^{th}$ subcarrier. $z[n]$ here also represents the input to the FFT at the receiver. Therefore, the output of the FFT can be expressed as

$$y[k] = \frac{1}{N} \sum_{n=0}^{N-1} z[n] \exp\left(\frac{j2\pi kn}{N}\right). \tag{4.3}$$

Substituting for $z[n]$ from (4.2) into (4.3) and after some algebraic manipulations, the output of the FFT is given by [41]

$$y[k] = \frac{1}{N} \sum_{m=0}^{N-1} s[m]H[m] \frac{\sin(\pi(m-k+\Delta f))}{\sin\left(\frac{\pi(m-k+\Delta f)}{N}\right)} \exp\left(j\left(\frac{N-1}{N}\right)(m-k+\Delta f)\right), \quad (4.4)$$

$$= \frac{1}{N} s[k]H[k]\left(\frac{\sin(\pi\Delta f)}{\sin(\pi\Delta f/N)}\right) \exp\left(\frac{j(N-1)\Delta f}{N}\right) + \frac{1}{N} \sum_{\substack{m=0 \\ m\neq k}}^{N-1} s[m]H[m]\beta_{m-k}, \quad (4.5)$$

where the complex coefficient

$$\beta_{m-k} = \frac{\sin(\pi(m-k+\Delta f))}{\sin\left(\frac{\pi(m-k+\Delta f)}{N}\right)} \exp\left(j\left(\frac{N-1}{N}\right)(m-k+\Delta f)\right). \quad (4.6)$$

Here, it can be seen that when $m = k$ we have

$$\beta_0 = \left(\frac{\sin(\pi\Delta f)}{\sin(\pi\Delta f/N)}\right) \exp\left(\frac{j(N-1)\Delta f}{N}\right), \quad (4.7)$$

which is identical to the scaling factor on the $k^{th}$ subcarrier in (4.5). This implies that in case of frequency offset, each output symbol estimate now depends on all the input values, i.e., ICI occurs due to the influence of data on the other subcarriers. Further, it can be seen from (4.5) that if $\Delta f = 0$ then the received signal is $s[k]H[k]/N$. Since the scaling of $k^{th}$ component is independent of $k$, it is evident that all subcarriers experience the same degree of attenuation along with ICI. It is important to note here that carrier frequency offset does not affect the amplitudes of any of the signals, and, consequently, it does not change the total power in the received signal. Therefore, the total ICI power changes little with $N$. Some techniques for offset estimation and offset cancelation are provided in the following section. More details can be found in [41, 42, 43].

## 4.2    FREQUENCY OFFSET ESTIMATION

By estimating the frequency offset at the receiver, the loss in performance due to a frequency mismatch of the received signal and the receive oscillator can be significantly reduced. The frequency offset estimation techniques can be broadly classified into pilot-aided schemes and non-pilot aided or blind estimation schemes. Pilot assisted methods use well defined pilot symbols to aid in the estimation of CFO. Since this method is capable of achieving very quick and reliable estimates, it is a popular technique though there is a loss in data rate and spectrum efficiency of the system. Blind or non pilot assisted methods exploit the structural and statistical properties of the transmitted OFDM signals. Though these techniques preserve the data rate, they lead to processing the received data multiple times, which causes delay in decoding. After normalizing the CFO by the subcarrier spacing, the integer part and the fractional part of the CFO can be estimated separately. Estimation of the integer part of the CFO can be termed as coarse CFO estimation while the estimation of the fractional part of the CFO can be termed as fine estimation of the CFO. Next, we describe briefly simple methods to estimate the integer part and the fractional part of the CFO.

## 4.2.1   FREQUENCY DOMAIN AUTOCORRELATION

For this method, pilot symbols are transmitted on a selected set of subcarriers. Out of $N$ subcarriers in an OFDM symbol, $J$ are selected to be pilots. These $J$ subcarriers are not necessarily contiguous. Since the integer part of the CFO causes frequency shift of the received signals in the frequency domain, this method yields good estimates of the CFO. Recall from Chapter 3 that an OFDM block consists of several OFDM symbols, and each OFDM symbol contains $N$ subcarriers, so that the data point $y[i, j]$ represents the symbol transmitted on the $j^{th}$ subcarrier of the $i^{th}$ OFDM symbol. For the frequency domain auto-correlation scheme, two consecutively received OFDM symbols on a set of subcarriers are correlated [44], as shown in Figure 4.2, to yield

$$\Delta \widehat{f}[g] = \sum_{j=0}^{J-1} y[i, \alpha_j + g] y^*[i - 1, \alpha_j + g], \qquad (4.8)$$

where $g = 0, \pm 1, \pm 2, \pm 3, \ldots$ are the possible integer-valued subcarrier shifts, and $\alpha_1, \alpha_2, \ldots, \alpha_J$ are the $J$ pilot subcarriers. Since the pilot symbols are not random but known at the receiver, (4.8) will contain the average magnitude of the squared pilot symbols. The integer portion of the CFO can be estimated by finding the value of $g$ which results in the largest $|\Delta \widehat{f}[g]|$, i.e.,

$$\widehat{g} = \underset{g}{\text{argmax}} \, |\Delta \widehat{f}[g]|. \qquad (4.9)$$

Pilot symbols have to be transmitted over several consecutive OFDM symbols to obtain a good estimate and minimize the error in estimation that maybe caused by channel fluctuations. Following this, to obtain an accurate estimate of the CFO, we describe the maximum likelihood method of estimating the fractional portion of CFO.

## 4.2.2   MAXIMUM LIKELIHOOD ESTIMATION

Though the cyclic-prefix can be used for timing and frequency synchronization, generally, in OFDM transmissions, there will be an additional preamble transmitted after the CP, and before the data is transmitted [37]. The preamble is designed to contain multiple repetitive symbols with a symbol time much less than that of the transmitted data symbol. Such a preamble can be used to estimate the fractional part of the CFO. Defining $Q$ as the repetition interval length in time samples and $B$ as the time samples separation between two adjacent repetitions, the maximum likelihood estimator can be expressed as

$$\Delta \widehat{f}_{frac} = \frac{1}{2\pi B T_s} \arg \left( \sum_{q=0}^{Q-1} z[n - q] z^*[n - q - B] \right), \qquad (4.10)$$

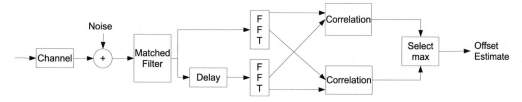

**Figure 4.2:** Block Diagram of Carrier Frequency Offset Estimation Process by using the Frequency Domain Approach.

where $\arg(\cdot)$ represents the argument of a complex number. Given that the phase can be uniquely resolved in the interval $[-\pi, \pi]$, the CFO can be estimated only within the interval $[-1/(2LT_s), 1/(2LT_s)]$. Adding this to the result obtained by the estimation of the integer part of CFO, a more accurate estimate is obtained.

Several other algorithms can be used for CFO estimation. In [41], the authors propose a correlation based technique for estimation. In this method, two consecutive identical pilot symbols are required to estimate CFO. The restricting assumption made is that the maximum CFO has to be less than half the subcarrier spacing. In [42], the authors used two identical half-period symbols to estimate the fractional part of the CFO and a second full period symbol that has a special correlation relation with the first pilot symbol to estimate the integer part of CFO. The important assumption the authors made in this work is that the constellation of symbols transmitted on each subcarrier has points that are equally spaced in phase. A similar method exploiting only two identical half period symbols to estimate both the integer and the fractional part of the CFO was proposed by [45]. While the above cited works depend on the correlation of the two half-period identical blocks for estimation, in [46, 47], the pilot symbol consists of multiple repetitive fractional parts. The differential phase of the correlation between different pairs of adjacent fractional blocks in a symbol are used to form an improved estimates.

In the blind estimation methods [37, 48, 49], elements of the transmitted OFDM symbol such as the cyclic prefix, virtual subcarriers or constant modulus transmission are used. Practical OFDM systems in general do not have data transmitted on all available subcarriers to help avoid aliasing errors. Some of the subcarriers at the edges of the OFDM symbol are left empty; these subcarriers are called virtual subcarriers. The number of subcarriers in a symbol is a system design parameter (generally about 10% of the total number of subcarriers $N$). The authors in [50] propose a blind estimation method that is only suitable to recover CFO values that are multiples of the subcarrier spacing. In [51, 48], the presence of virtual subcarriers is exploited and techniques such as MUSIC and ESPIRIT [26] are used to estimate the CFO. This scheme requires usually multiple OFDM symbols to achieve desirable performance thereby leading to additional delay at the receiver to estimate the CFO and decode the received symbols.

In a typical communication system, offset estimation is done in the presence of channel noise corrupting the received signals. Therefore, the estimates obtained are always noisy. When these estimates are used to reverse the effects of the frequency offset, there is a residual offset that is small, but random. This results in deterioration of performance, in spite of compensating for carrier offset using the estimation process. Therefore, it is preferable that carrier offset be canceled automatically, rather than be estimated and then removed. In Section 4.3, some algorithms for ICI cancelation are presented.

## 4.3   ICI CANCELATION SCHEMES

### 4.3.1   SELF-ICI CANCELATION SCHEME

There have been several schemes proposed to avoid ICI in the OFDM communication scheme. The first scheme we consider is called *Self ICI Cancelation*, proposed by Zhao and Haggman [52, 53]. In this scheme, instead of independent data being mapped on to the subcarriers, data is mapped onto adjacent pairs of subcarriers. For example, $s[0] = -s[1], s[2] = -s[3], \ldots, s_[N-2] = s_[N-1]$. This mapping has been shown to result in cancelation of most of the ICI in the values $y[0], \ldots, y[N-1]$. So, it is evident that the ICI for this scheme depends on the difference between the adjacent weighting coefficients rather than on the coefficients themselves. As the difference between adjacent subcarriers is small this results in substantial reduction in ICI. If adjacent coefficients are identical, then ICI is completely canceled. The ICI cancelation in this scheme depends only on the coefficients being slowly varying functions of offset, and it does not depend on the absolute value of the coefficients themselves. However, due to the redundancy introduced by mapping the same symbol onto two subcarriers, the data rate is halved.

### 4.3.2   WINDOWING

Windowing is another technique proposed to help reduce sensitivity to frequency offsets in an OFDM system [54, 55, 56, 57]. This process involves cyclically extending the time domain signal associated with each symbol by $v$ samples. The resulting signal is then shaped with a window function.

The transmitter uses an $N/2$ point IDFT process while the receiver uses an $N$ point DFT process. If the time domain signal is extended by $v = N/2$ samples, then $N$ point received signal can be used as inputs to the DFT process at the receiver. If $v < N/2$ then zero padding can be employed to obtain a sequence of length $N$. At the output of the DFT process, the even numbered outputs are used to estimate the transmitted symbols while the odd numbered outputs are discarded. Here it is important to note that since all the received power is not being used in generating data estimates, this method has a reduced overall SNR compared with OFDM without windowing. Different windows can be used in this scheme. The authors in [54] consider a Hanning window, while in [55] the general class of windows satisfying the Nyquist criteria are studied, and the Kaiser window is studied in [55]. Details about these and other windowing techniques can be seen in [58].

Though the self ICI cancelation method is simple to implement, each data symbol is carried on two sub-carriers. Therefore, the data rate of the system and the frequency efficiency of the system is reduced by half. In the windowing method, the functions used to cancel ICI have non-zero side bands, leading to the addition of spurious bits and causing a loss in SNR. Thus, the choice of canceling scheme leads to a trade-off between data rate and SNR, which is dictated by the system design [43].

CHAPTER 5

# Peak to Average Power Ratio

While the carrier frequency offset is a phenomenon that occurs due to frequency mismatch at the receiver, high peak-to-average power ratio occurs at the transmitter due to summation of multiple sinusoids. Occasionally, these sinusoids can add coherently to yield a very high amplitude compared to the average amplitude, resulting in a large peak-to-average power ratio (PAPR). To ensure that these peaks are transmitted without distortion, the power amplifier at the transmitter should be capable of remaining linear over a wide range of input amplitudes. This presents a significant challenge in terms of design, cost and power consumption. In this chapter, we describe this occurrence in further detail and present schemes to minimize the effects of a high PAPR.

## 5.1 PROBLEM FORMULATION

An OFDM signal consists of a number of independently modulated SCs, which can result in a large PAPR when added up coherently. The different carriers may align in phase at some instant in time, and, therefore, they produce an amplitude peak equal to the sum of the amplitudes of the individual carriers. This occurs with extremely low probability for large N.

The peak power is defined as the power of a sine wave with an amplitude equal to the maximum envelope value. Hence, an unmodulated carrier has a PAPR of 0 dB. An alternative measure of the envelope variation of a signal is the crest factor, which is defined as the maximum signal value divided by the RMS signal value. For an unmodulated sinusoidal carrier, the crest factor is 3 dB. This 3 dB difference between the PAPR and crest factor also holds for other non-sinusoidal carriers, provided that the center frequency is large in comparison with the signal bandwidth. A large PAPR has disadvantages like a requirement of increased complexity of analog-to-digital (A/D) and digital-to-analog (D/A) converters, and reduced efficiency of the RF power amplifier.

The output of the IFFT at the transmitter can be represented as

$$x[n] = \frac{1}{N} \sum_{k=0}^{N-1} s[k] \exp\left(\frac{j2\pi kn}{N}\right). \tag{5.1}$$

Using this, the peak power of transmission can be expressed as

$$\max_n \left\{ |x[n]|^2 \right\} = \frac{1}{N^2} \max_n \left\{ \sum_{k_1=0}^{N-1} \sum_{k_2=0}^{N-1} s[k_1] s^*[k_2] \exp\left(\frac{j2\pi(k_1 - k_2)n}{N}\right) \right\}, \tag{5.2}$$

where $|x[n]|^2 = x[n]x^*[n]$. Similarly, the average power can be expressed as

$$\mathrm{E}\left[|x[n]|^2\right] = \frac{1}{N^2}\mathrm{E}\left[\sum_{k_1=0}^{N-1}\sum_{k_2=0}^{N-1} s[k_1]s^*[k_2]\exp\left(\frac{j2\pi(k_1 - k_2)n}{N}\right)\right]. \tag{5.3}$$

The PAPR, using (5.2), and (5.3), can be expressed as

$$\mathrm{PAPR} = \frac{\max_{n}\left\{|x[n]|^2\right\}}{\mathrm{E}\left[|x[n]|^2\right]}. \tag{5.4}$$

As an example, consider the case when BPSK modulation is used, i.e., $s[k] \in \{-1, 1\}$. For this case, from (5.2), the peak transmit power is one, and $\mathrm{E}\left[|x[n]|^2\right] = 1/N$, thereby leading to a PAPR of $N$. For example, for an OFDM system employing 1024 subcarriers per transmitted symbol, the PAPR$= 1024 \approx 30$dB, which is an extremely large range for the transmit power amplifier to vary over.

The techniques proposed for PAPR reduction can be divided into three categories: signal distortion techniques, coding techniques, and scrambling techniques. In signal distortion techniques, nonlinear distortion is introduced in the OFDM signal at or around the peaks. Examples of distortion techniques include clipping, peak windowing, and peak cancelation. Coding techniques use forward error correcting codes that exclude OFDM symbols with a large PAPR. In scrambling techniques, each OFDM symbol is scrambled with a different scrambling sequence, which is selected to yield the smallest PAPR.

An example is shown in Figure 5.1 where several OFDM subcarriers are added together, as transmitted. This leads to some peaks forming that are high, and consequently, a high PAPR. In the example shown, it is assumed that an amplitude greater than a predefined threshold causes a PAPR that is unacceptable. In the time section shown, there is one such peak.

## 5.2   PAPR MITIGATION METHODS

### 5.2.1   SIGNAL DISTORTION TECHNIQUES

Since large PAPR occurs rarely, the peaks can be removed at the cost of a slight amount of self-interference. The simplest way to remove the peaks is by clipping the signal such that the peak amplitude becomes limited to some predefined maximum level. By defining the highest accepted peak value as the *clipping threshold*, any peak above this value will be clipped appropriately. Since clipping can be viewed as a rectangular windowing operation in time, non-linear distortion introduced by clipping, called self-interference, causes deterioration of the error rate performance of the system and also significantly increases the out-of-band radiation levels. Due to the slow roll-off of the spectrum of the rectangular window and the large side-lobes, the out-of-band radiation levels are high. Different window shapes, other than rectangular, have been considered to minimize the out-of-band radiation level, including the Gaussian, raised cosine, Kaiser and Hamming windows. To

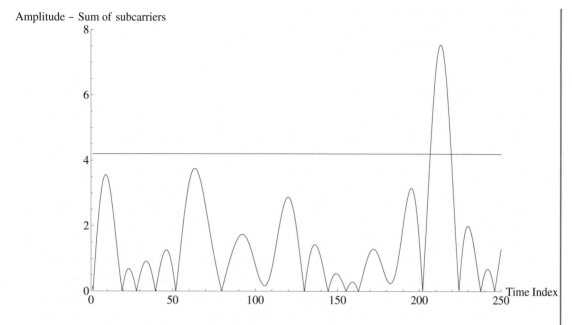

**Figure 5.1:** Amplitude of transmitted OFDM symbol.

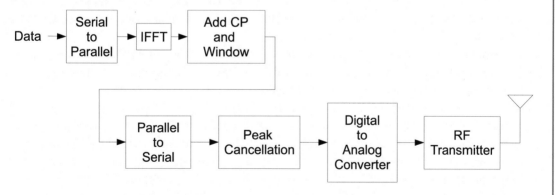

**Figure 5.2:** PAPR reduction by Peak Cancelation.

minimize the out-of-band interference, ideally, the window should be narrow in frequency and have a fast roll-off with small side-lobes.

Peak cancelation can also be performed digitally. A comparator is used to check if the peak amplitude of the digital OFDM symbol is above a predefined threshold, and if it is above the threshold, the peak and the side lobes are scaled appropriately to maintain the PAPR to a predefined value. Figure 5.2 shows the block diagram of an OFDM transmitter implementing peak cancelation. As shown in Figure 5.2, the peak cancelation procedure is performed after the addition of the CP.

Peak cancelation can also be performed on a symbol-by-symbol basis immediately after the IDFT, before adding the cyclic prefix and windowing. There is no change needed in the receiver architecture for the digital peak cancelation technique.

## 5.2.2   CODING AND SCRAMBLING

Though peak cancelation offers a simple yet powerful technique to control the PAPR of an OFDM system, an important drawback of this technique is that symbols with a large PAPR suffer more degradation, so they are more vulnerable to errors. Given that the PAPR is high only once in several OFDM symbols, another technique to minimize the effects of PAPR is error control coding. By using codes with low rates, i.e., with high redundancy, errors caused by symbols with a large degradation can be corrected by the surrounding symbols. The authors in [59], by exhaustively searching all possible QPSK code words, have shown that for eight channels, a rate 3/4 convolution code exists that provides a maximum PAPR of 3 dB. Also, in [59], it is illustrated that many of the codes developed for PAPR reduction are Golay complementary sequences. Golay complementary sequences are sequence pairs for which the sum of autocorrelation functions is zero for all delay shifts not equal to zero [60, 61, 62]. In [63], the author presents a specific subset of Golay codes, together with decoding techniques that combine PAPR reduction with good error correcting capabilities. But, if the received signal is suffering from burst errors, then the initial transmission and the retransmission might both have a large number of errors even with coding. To deal with this, scrambling techniques are used to ensure that the transmitted data between initial transmission and retransmissions are uncorrelated.

Symbol scrambling techniques to reduce the PAPR of a transmitted OFDM signal can be seen as a special type of a PAPR reduction code. Symbol scrambling does not, however, try to combine error correcting coding and PAPR reduction such as is done by complementary codes. The basic idea of symbol scrambling is that for each OFDM symbol, the input sequence is permuted by a set of scrambling sequences and the output signal with the smallest PAPR is transmitted. For uncorrelated scrambling sequences, the resulting OFDM signals and corresponding PAPRs will be uncorrelated, so if the PAPR for one OFDM symbol has a probability $p$ of exceeding a certain level without scrambling, the probability is decreased to $p^K$ by using $K$ scrambling codes. Hence, symbol scrambling does not guarantee a PAPR below some low level; rather, it decreases the probability that high PAPRs will occur. Scrambling techniques were first proposed in [64] under the names selected mapping and partial transmit sequences. The difference between the two is that the first applies independent scrambling permutations to all SCs, while the latter only scrambles groups of SCs.

Though all three of the above mentioned methods help reduce the PAPR of an OFDM system [28], they each have drawbacks. While the coding method introduces redundancy and thereby a loss in transmission data rate, clipping of the peak amplitude introduces non-linear distortion into the system, and the scrambling method increases the complexity of the system and also the transmission overhead due to the need to transmit the scrambling sequence resulting in the lower

PAPR. Therefore, system design requirements are used to decide which of these schemes is used to overcome the effects of high PAPR on the system.

CHAPTER 6

# Simulation of the Performance of OFDM Systems

## 6.1 PERFORMANCE OF AN OFDM SYSTEM

In the performance analysis of an OFDM system, we assume that the channel remains constant for a certain length of time and then randomly changes to an independent value. This behavior is termed as quasi-static fading. Assuming quasi-static channel and perfect synchronization at the receiver leads to the received signals on the various subcarriers (SC) to be independent of each other. Therefore, the channel on each SC can be equivalently represented as a flat fading channel with additive white Gaussian noise (AWGN). The instantaneous signal to noise ratio (SNR) on each subcarrier, within a block of quasi-static fading, can now be represented as

$$\gamma_k = \bar{\gamma}|H[k]|^2, \tag{6.1}$$

where $\bar{\gamma}$ represents the average SNR on the subcarrier and $H[k]$ represents the channel on the $k^{th}$ subcarrier as given in (3.12). Note here that we have assumed that all SC's have the same average SNR $\bar{\gamma} = E_b/N_o$, where $E_b$ represents the average energy per bit and $N_o$ represents the height of the noise spectral density expressed in the units of Watts per Hertz.

Further, given that the subcarriers are i.i.d., and assuming perfect channel knowledge at the receiver, the probability of error, $P_e$, of the OFDM system can be expressed as the mean of the probability of error of individual subcarriers, i.e.,

$$P_e = \frac{1}{N}\sum_{k=1}^{N} P_e[k], \tag{6.2}$$

where $P_e[k]$ is the channel dependent instantaneous probability of error on the $k^{th}$ subcarrier. $P_e[k]$ depends on the modulation scheme chosen. For binary phase shift keying (BPSK) modulation [65, Chap. 3], we have

$$P_e[k] = Q\left(\sqrt{2\gamma_k}\right). \tag{6.3}$$

If instead, quadrature phase shift keying (QPSK) modulation is used, we have

$$P_e[k] = 2Q\left(\sqrt{\gamma_k}\right) - Q^2\left(\sqrt{\gamma_k}\right). \tag{6.4}$$

For other modulation schemes, instantaneous probability of error expressions can be found in [19].

To average the probability of error over time, the instantaneous value has to be averaged across all possible values of the random variable $H[k]$, i.e.,

$$\overline{P}_e[k] = E_H\left[P_e[k]\right], \tag{6.5}$$

where $E_H[\cdot]$ denotes the expectation operator with respect to $H[k]$. Therefore, using (6.5), the average probability of error of a OFDM system can be expressed as

$$\overline{P}_e = \frac{1}{N}\sum_{k=1}^{N} E_H\left[P_e[k]\right]. \tag{6.6}$$

At the receiver, a maximum likelihood (ML) decoder is implemented. An ML decoder, as the name implies, maximizes the likelihood of receiving a signal, $\mathbf{y}$, conditioned on the transmitted signal, $\mathbf{s}$, and channel, $\mathbf{H}$, i.e.,

$$\hat{\mathbf{s}} = \underset{\mathbf{s}}{\operatorname{argmin}}\, p(\mathbf{y}|\mathbf{s}, \mathbf{H}) = \underset{\{s_k\}}{\operatorname{argmax}} \prod_{k=1}^{N} p(y_k|s_k, h_k). \tag{6.7}$$

For all simulations considered, we assume Rayleigh fading channels and additive white Gaussian noise (AWGN), thereby leading to an ML decoder of the form [19],

$$\hat{\mathbf{s}} = \underset{\mathbf{s}}{\operatorname{argmax}} \|\mathbf{y} - \mathbf{H}\mathbf{s}\|^2. \tag{6.8}$$

For the case when frequency and timing synchronization is perfect at the receiver, ML decoder can be expressed as

$$\hat{s}[k] = \underset{s[k]}{\operatorname{argmin}} |y[k] - H[k]s[k]|^2. \tag{6.9}$$

Note here that the ML decoder can be represented as in (6.9) because there is no channel coding. Instead, if channel coding was employed then sequence detection algorithms like the Viterbi algorithm [17] needs be used.

In the following section, we implement the Monte-Carlo method to calculate the average probability of error of an OFDM system. The Monte-Carlo method is a numerical method to estimate the ensemble average with respect to a random variable. Further information on how this technique can be used to approximate the expected value can be found in [66, 67, 68, 69, 70].

## 6.2   SIMULATIONS

In what follows, MATLAB simulations are used to demonstrate the working of OFDM, based on the theoretical development of OFDM has been presented previously.

Figure 6.1: Basic OFDM model.

## 6.2.1 THE BASIC OFDM SYSTEM

In this section, the basic OFDM system is simulated. As discussed in the preceding chapters, a basic OFDM system is constructed as shown in Figure 6.1. In this section, MATLAB code will be provided for each of the blocks, with an explanation. The full program for the basic OFDM system is then provided, including flexibility to vary several parameters and visualization options.

**Data Generation and Modulation**

Data is first generated to be transmitted over the OFDM system. Data of length N is randomly generated and modulated as shown:

```
am = [-1,1];
M = 2;
dat_ind = ceil(M*rand(1,N));
data = am(dat_ind);
```

These lines of code generate a baseband representation of BPSK signals (±1). The data is generated from a source that generates the symbols with equal probability. It should be noted here that BPSK is not the only mode of modulation, and any other modulation scheme can be used at this stage. For example, if QPSK is used, the code can be modified as follows:

```
am = [1,1i,-1,-1i];
M = 4;
dat_ind = ceil(M*rand(1,N));
data = am(dat_ind);
```

**IFFT**

The first operation performed on the data is a N-point IFFT. In MATLAB, the IFFT function can be used for this operation as follows:

```
data_t = ifft(data);
```

**Add Cyclic Prefix**

After the IFFT, the last few data-points are repeated at the beginning. The repeated data is called the cyclic prefix.

```
data_cp = [data_t(end-CP_length+1:end), data_t];
```

Cyclic prefix of length CP_length is added to the beginning of the data block.

## Channel

Transmission occurs over frequency selective fading channels. The channels are modeled as FIR filters of order $L$. The channel, $h$, is assumed to have channel taps drawn from a Rayleigh distribution. The power per tap is normalized, and the channel can be simulated as

```
h = complex(randn(L+1,1), randn(L+1,1))*sqrt(0.5/(L+1));
```

Noise is generated to be added to the transmission over the channel. The noise is zero-mean complex Gaussian. The code below shows how to generate the noise:

```
noise = complex(randn(1,Total_length), randn(1,Total_length))*sqrt(0.5/N);
```

Since the channel is modeled as an FIR filter, the output of the channel is computed by filtering the input signal with the channel, and noise is then added on to it. To adjust the SNR, the transmission is scaled by a power value as shown:

```
rho = SNR;
rec = sqrt(rho)*(filter(h,1,data_cp))+noise
```

## Remove Cyclic Prefix

The cyclic prefix is removed from the received data. The first CP_length symbols are discarded from the received data:

```
rec_sans_cp = rec(CP_length+1:end)
```

## FFT and Demodulation

The data extracted by discarding the cyclic prefix is transformed into the frequency domain. The data in the frequency domain is then equalized to account for the channel, and to yield the final received signal. It should be noted here that the FFT used here to convert the channel into the frequency domain is normalized by the number of channel taps. The received signal is then demodulated to obtain an estimate of the transmitted signals. In the following code, it is assumed that BPSK modulation has been used at the transmitter:

```
rec_f = fft(rec_sans_cp); % FFT
h_f = sqrt(rho)*fft(h,N); % Equivalent channel on each subcarrier
det1 = abs(rec_f+h_f).^2; % Calc the Euclidean dist
                          % assuming -1 was transmitted
det2 = abs(rec_f-h_f).^2; % Calc the Euclidean dist
                          % assuming +1 was transmitted
det = [det1, det2];       % Concatenating the two vectors

% Find the symbol the received signal is closest to
[min_val, ind] = min(det, [], 2);
```

```
dec = 2*((ind-1)>0.5)-1;  % BPSK decoding
```

For QPSK decoding:

```
rec_f = fft(rec_sans_cp); % FFT

% Equivalent channel on each subcarrier
h_f = sqrt(rho)*fft(h,N);

% Calc the Euclidean dist assuming 1 was transmitted
det1 = abs(rec_f-am(1)*h_f).^2;

% Calc the Euclidean dist assuming +1i was transmitted
det2 = abs(rec_f-am(2)*h_f).^2;

% Calc the Euclidean dist assuming -1 was transmitted
det3 = abs(rec_f-am(3)*h_f).^2;

% Calc the Euclidean dist assuming -1i was transmitted
det4 = abs(rec_f-am(4)*h_f).^2;

% Concatenating the vectors
det = [det1, det2, det3, det4];

% Find the symbol the received signal is closest to
[min_val, ind] = min(det, [], 2);

% Generating the decoded symbols
dec = am(ind);
```

The received symbols are decoded using maximum likelihood (ML) estimation.

### 6.2.1.1  OFDM Simulation

A Monte-Carlo simulation is performed to estimate the probability of error of the received signals when transmission occurs using OFDM. The SNR of the channels is varied and the probability of error is estimated for each case of SNR. The program for the full simulation model is provided. In the code provided, it is assumed that data is transmitted in frames, and each frame consists of B OFDM symbols. This allows us to evaluate the effects of inter-block-interference when the length of the cyclic prefix is less than the number of channel taps. The simulation results are presented for different parameters, which can be changed within the code.

```
clear all
clc

N = 16;              % Number of subcarriers in each OFDM symbol
L = 3;               % Channel order
CP_length = 4;       % Cyclic prefix length
B = 10;              % Number of OFDM symbols per transmitted frame
mc_N = 5000;         % Number of iterations to achieve sufficient errors
SNR_db = 0:2:20;                      %SNR in dB
SNR = 10.^(SNR_db/10);                % SNR values
Pe = zeros(size(SNR_db));             % Initializing the error vector
Total_length = (CP_length+N)*B;       % Total length of each frame
am = [-1,1];         % For BPSK
M = 2;               % For BPSK

for SNR_loop = 1:length(SNR_db)
    rho = SNR(SNR_loop);
    err = 0;

    for mc_loop = 1:mc_N
        dat_ind = ceil(M*rand(B,N));
        data = am(dat_ind);

        % Reshaping the data into a BxN matrix,...
        %...used later for error detection
        data_reshape = reshape(data, 1, B*N);

        tx_data = data;

        for b = 1:B
            % Taking the IFFT
            data_t(b,:) = ifft(tx_data(b,:));
        end

        % Adding Cyclic prefix
        data_cp = [data_t(:,end-CP_length+1:end), data_t];

        % Reshape the BxN matrix to obtain the frame (1xTotal_length)
        data_tx =reshape(data_cp.',1,Total_length);
```

```
h = complex(randn(L+1,1), randn(L+1,1))*sqrt(0.5/(L+1));

%Noise
noise = complex(randn(1,Total_length), ...
                    randn(1,Total_length))  * sqrt(0.5/N);
% Received signal
rec = sqrt(rho)*(filter(h,1,data_tx))+noise;

% Reshape the recd signal into CP_length+N x B array
rec_reshaped = (reshape(rec, CP_length+N, B)).';

% Remove CP
rec_sans_cp = rec_reshaped(:,CP_length+1:end);

for bb = 1:B
    % Taking the FFT
    rec_f(bb,:) = fft(rec_sans_cp(bb,:));
end

% Calculating the equivalent channel on each subcarrier
h_f = sqrt(rho)*fft(h,N);

for b2 = 1:B

    % Extracting the OFDM symbol from the "rec_f" matrix
    rec_symbol = transpose(rec_f(b2,:));

    % Calc Euclidean dist assuming -1
    det1 = abs(rec_symbol+h_f).^2;

    % Calc Euclidean dist assuming +1
    det2 = abs(rec_symbol-h_f).^2;

    % Concatenate the two vectors
    det = [det1, det2];

    % Find symbol the recd signal is closest to
    [min_val, ind] = min(det, [], 2);
```

```
            % Generate the decoded symbols
            dec(b2,:) = 2*((ind-1)>0.5)-1;
        end

        % Reshape the decoded symbols to calc error
        dec_reshape = reshape(dec, 1, B*N);

        % Comparing dec_reshape against...
        %...data_reshape to calculate errors
        err = err + sum(dec_reshape~=data_reshape);
    end
    % Calculate the probability of error
    Pe(SNR_loop) = err/(mc_N*B*N);
end

% Semilog plot of Pe vs. SNR_db
semilogy(SNR_db,Pe)
```

Figure 6.2 shows the output of the code for two modulation schemes, BPSK and QPSK. It is shown that the performance of the system remains the same irrespective of the number of subcarriers, as long as the number of symbols in the cyclic prefix is at least as many as the number of channel taps. Furthermore, as expected from digital communications, the symbol error rate of the QPSK system is worse than the BPSK case. In Figure 6.3, the system is simulated for QPSK modulation. Three curves are plotted, each for a different length of cyclic prefix. It can be seen that as the length of the cyclic prefix reduces, inter-block-interference (IBI) causes deterioration in performance.

## 6.2.2    CARRIER FREQUENCY OFFSET

Carrier frequency offset and its drawbacks were discussed in Chapter 4. In this Section, we will provide simple code to simulate the effect of frequency offset and demonstrate its effect on the BER performance of the OFDM system. More detailed treatment can be found in [71, 72].

### 6.2.2.1  Simulation

The frequency offset is provided as a complex exponential multiplier for each subcarrier. This simulates the effect of having the frequency of each subcarrier being offset at the receiver by a small amount. In the code that follows, the frequency offset is varied, and the BER is calculated for each of those values, to demonstrate the effect of CFO on performance of OFDM systems. Similar to the previous cade, the code is flexible and different parameters such as modulating techniques, levels of offset and number of Monte-Carlo iterations can be changed.

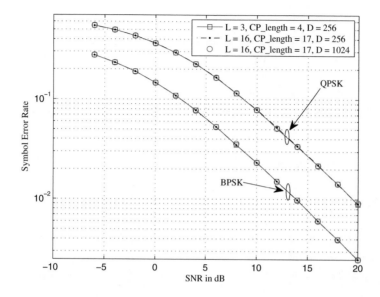

**Figure 6.2:** Simulation of the basic OFDM model. Plot shows effect of SNR on the probability of error.

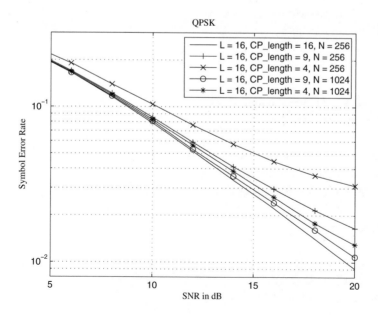

**Figure 6.3:** Simulation of the basic OFDM model, showing the need for cyclic prefix.

```matlab
clear all
clc

N = 16;         % Number of subcarriers in each OFDM symbol
L = 3;          % Channel order
CP_length = 4;  % Cyclic prefix length
B = 1;          % Number of OFDM symbols per transmitted frame
mc_N = 5000;    % Number of iterations to achieve sufficient errors
SNR_db = 5;     % SNR in dB


SNR = 10.^(SNR_db/10); % SNR


Pe = zeros(size(SNR_db));


Total_length = (CP_length+N)*B; % Total length of each frame
am = [-1,1];    % For BPSK
M = 2;          % For BPSK


freq_offset = [-0.5:0.01:0.5];

for off_loop = 1:length(freq_offset)
    rho = SNR;
    err = 0;
    for mc_loop = 1:mc_N

        dat_ind = ceil(M*rand(B,N));
        data = am(dat_ind);

        % Reshaping the data into a BxN matrix,...
        %...used later for error detection
        data_reshape = reshape(data, 1, B*N);

        tx_data = data;

        for b = 1:B
            % Taking the IFFT
            data_t(b,:) = ifft(tx_data(b,:));
        end
```

```
% Adding Cyclic prefix
data_cp = [data_t(:,end-CP_length+1:end), data_t];

% Reshape the BxN matrix to...
%...obtain the frame (1xTotal_length)
data_tx =reshape(data_cp.',1,Total_length);
h = complex(randn(L+1,1), randn(L+1,1))*sqrt(0.5/(L+1));
noise = complex(randn(1,Total_length), ...
                    randn(1,Total_length)) *sqrt(0.5/N); %Noise
rec = sqrt(rho)*(filter(h,1,data_tx))...
                * exp(-1i*2*pi*freq_offset(off_loop))+noise;

% Reshape the recd signal...
%...into CP_length+N x B array
rec_reshaped = (reshape(rec, CP_length+N, B)).';

% Remove CP
rec_sans_cp = rec_reshaped(:,CP_length+1:end);

for bb = 1:B

    % Taking the FFT
    rec_f(bb,:) = fft(rec_sans_cp(bb,:));
end

% Calculating the equivalent channel on each subcarrier
h_f = sqrt(rho)*fft(h,N);

for b2 = 1:B

    % Extracting the OFDM symbol from...
    %...the "rec_f" matrix
    rec_symbol = transpose(rec_f(b2,:));

    % Calc Euclidean dist assuming -1
    det1 = abs(rec_symbol+h_f).^2;

    % Calc Euclidean dist assuming +1
    det2 = abs(rec_symbol-h_f).^2;
```

```
            % Concatenate the two vectors
            det = [det1, det2];

            % Find symbol the recd signal is closest to
            [min_val, ind] = min(det, [], 2);

            % Generate the decoded symbols
            dec(b2,:) = 2*((ind-1)>0.5)-1;
        end

        % Reshape the decoded symbols to calc error
        dec_reshape = reshape(dec, 1, B*N);

        % Compare dec_reshape against...
        %...data_reshape to calc errors
        err = err + sum(dec_reshape~=data_reshape);
    end

    % Calculating the probability of error
    Pe(off_loop) = err/(mc_N*B*N)
end

%Semilog plot of Pe vs. offset
semilogy(freq_offset,Pe)
```

The code provided yields a value for the probability of error at each value of frequency offset. The value of probability of error is plotted for each value of offset. As expected, the best performance is obtained for a zero offset. As the offset increases in either direction, the performance deteriorates, and the performance is symmetric in the offset about the zero offset point. This plot for a channel SNR of 5dB is shown in Figure 6.4.

### 6.2.3   PAPR SIMULATIONS

The effect of high peak-to-average-power ratio (PAPR) on the performance of OFDM systems was seen in Chapter 4. In this simulation, the transmitted signal is simulated to provide an example of the transmit power over a few time samples and to demonstrate the occurrence of high PAPR. In this simulation, the base-band equivalent model of an OFDM system is considered. At each time instant, data is BPSK modulated and transmitted using OFDM. The instantaneous peak transmit power and the average transmit power are calculated and plotted in Figure 6.5. The number of subcarriers used is 16. It can be verified from (5.3) that the average power of the system must be

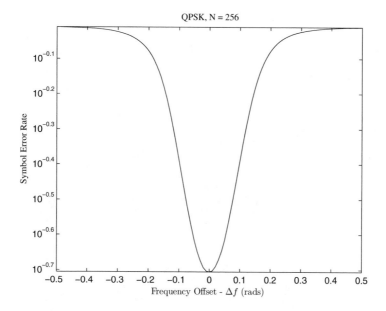

Figure 6.4: Bit error rate vs. phase offset

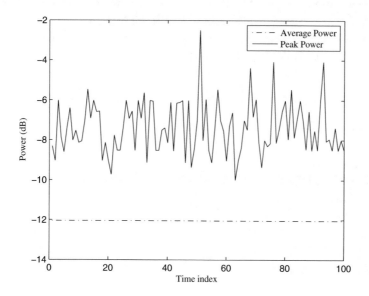

Figure 6.5: Peak power due to addition of in-phase sinusoids, compared with the average power of the transmissions.

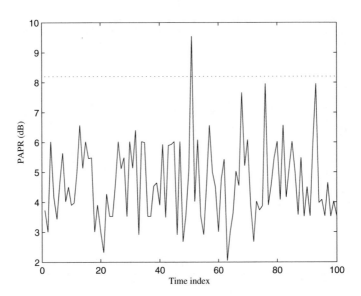

**Figure 6.6:** Peak to average power ratio compared against a threshold.

-12dB as can be seen from the figure. The PAPR is also calculated and plotted in Figure 6.6. As can be seen from (5.4), the maximum possible value of PAPR is 12dB, and in the case of this simulation, the system does not exceed this value. In fact, the largest PAPR in this case approaches 10dB. A predetermined threshold is also shown, and the PAPR exceeds this threshold at one point.

```
clear all

N = 16; % Length of data
SNR_db = 5;
SNR = 10.^(SNR_db/10);
Pe = zeros(size(SNR_db));
noise_var = 1/SNR;
time_samples = 100;
avg_pow = zeros(1,time_samples);
mx_pow = zeros(1,time_samples);
papr = zeros(1,time_samples);

for time_loop = 1:time_samples
    data = 2*(randn(N,1)>0)-1;
    data_t = ifft(fftshift(data));
```

```
    avg_pow(time_loop) = (norm(data_t))^2/N;
    mx_pow(time_loop) = max(data_t.*conj(data_t));
    papr(time_loop) = mx_pow(time_loop)/avg_pow(time_loop);
end

figure(1)
plot(10*log10(avg_pow))
hold all
plot(10*log10(mx_pow))
figure(2)
plot(10*log10(papr))
```

One of the ways of limiting PAPR at the transmitter is by clipping signals that exceed a certain level. In the code provided, an OFDM system is considered with such a clipping systems. Different clipping thresholds are considered to show the effect of clipping level on performance. In this example, the system considered has no additive noise, and the channel is frequency flat with gain one.

```
clear all
clc


N = 16;           % Number of subcarriers in each OFDM symbol
L = 3;            % Channel order
CP_length = 4;    % Cyclic prefix length
B = 10;           % Number of OFDM symbols per transmitted frame
mc_N = 50;        % Number of iterations to achieve sufficient errors
th_var = 0:.1:1;  % Clipping thresholds
Pe = zeros(size(th_var));      % Initializing the error vector
Total_length = (CP_length+N)*B; % Total length of each frame
am = [1,1i,-1,-1i];            % For QPSK
M = 4;                         % For QPSK

for th_loop = 1:length(th_var)
    th = th_var(th_loop);
    err = 0;

    for mc_loop = 1:mc_N
        dat_ind = ceil(M*rand(B,N));
        data = am(dat_ind);

        % Reshape the data into a BxN matrix,...
```

```matlab
%...used later for error detection
data_reshape = reshape(data, 1, B*N);
tx_data = data;

for b = 1:B
    % Taking the IFFT
    data_t(b,:) = ifft(tx_data(b,:));
end

% Adding Cyclic prefix
data_cp = [data_t(:,end-CP_length+1:end), data_t];

% Reshaping the BxN matrix to...
%...obtain the frame (1xTotal_length)
data_tx =reshape(data_cp.',1,Total_length);

thu = abs(th);
thl = -abs(th);
data_clip = data_tx;
pt_high = find(data_clip>thu);
data_clip(pt_high) = thu;

pt_low = find(data_clip<thl);
data_clip(pt_low) = thl;
rec = data_clip;

% Reshape the recd signal...
%...into CP_length+N x B array
rec_reshaped = (reshape(rec, CP_length+N, B)).';

% Remove CP
rec_sans_cp = rec_reshaped(:,CP_length+1:end);

for bb = 1:B

    %Taking the FFT
    rec_f(bb,:) = fft(rec_sans_cp(bb,:));
end
```

```
        for b2 = 1:B
            % Extracting the OFDM symbol...
            %...from the "rec_f" matrix
            rec_symbol = transpose(rec_f(b2,:));

            % Calc the Euclidean dist assuming 1
            det1 = abs(rec_symbol-am(1)).^2;

            % Calc the Euclidean dist assuming +1i
            det2 = abs(rec_symbol-am(2)).^2;

            % Calc the Euclidean dist assuming -1
            det3 = abs(rec_symbol-am(3)).^2;

            % Calc the Euclidean dist assuming -1i
            det4 = abs(rec_symbol-am(4)).^2;

            % Concatenating the vectors
            det = [det1, det2, det3, det4];

            % Find symbol the recd signal is closest to
            [min_val, ind] = min(det, [], 2);

            % Generating the decoded symbols
            dec(b2,:) = am(ind);
        end
        % Reshape decoded symbols to calc error
        dec_reshape = reshape(dec, 1, B*N);

        % Compare dec_reshape against data_reshape to calculate errors
        err = err + sum(dec_reshape~=data_reshape);
    end
    % Calculate the probability of error
    Pe(th_loop) = err/(mc_N*B*N);
end

% Semilog plot of Pe vs. clipping threshold
semilogy(th_var,Pe)
```

Figure 6.7 shows the effect of clipping on the performance of an OFDM system. As expected, as the clipping threshold becomes smaller, the performance of the system deteriorates even though it is a noiseless, unit-gain frequency-flat channel. This is the trade-off between performance and power consumption. In order to keep the power consumption of the amplifier at the transmitter low, the clipping threshold is lowered, and this leads to poor performance. On the other hand, if the clipping threshold increases, the BER is lowered, but the power consumption increases.

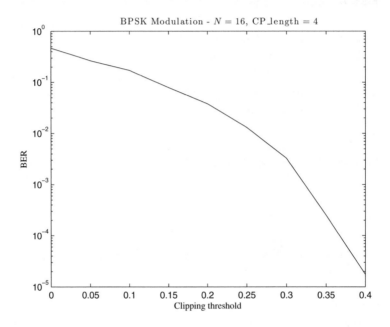

**Figure 6.7:** Effect of clipping on the BER of an OFDM system.

# CHAPTER 7

# Conclusions

This book examined and analyzed various aspects of Orthogonal Frequency Division Multiplexing (OFDM), including the use of OFDM in various current standards that demand high data rates and very low error rates when transmitting over wireless multipath channels. Challenges to high data rate and low error rate transmissions are analyzed by examining the characteristics of wireless communication channels. It is known that a frequency selective fading model best fits a wireless multipath channel. Equalization of the multipath channel in the frequency domain instead of the time domain is presented, and it is exploited in the design of multicarrier systems. A simple frequency domain multiplexing scheme is first described as a possible solution for communication over frequency selective channels. The drawbacks of this method are used to motivate the development of OFDM, which is then described in detail.

Using the FFT as a means for transformation from the time domain to the frequency domain, a discrete time OFDM baseband system can be easily developed. Since implementation of FFT is inexpensive in terms of simplicity and cost, and efficient in terms of computation speed, OFDM has become a popular choice for communication over frequency selective fading channels. An OFDM system is designed to have orthogonal subcarriers, and each subcarrier sees a flat fading channel. While this simplicity is an obvious advantage, OFDM does have weaknesses. Two main pathologies of the OFDM communication scheme, carrier frequency offset (CFO) and high peak to average power ratio (PAPR), are presented. When there is an offset in frequency between the carrier and the local oscillator at the receiver, the subcarriers do not remain orthogonal, leading to ICI. A high PAPR of the signal at the transmitter makes the design of efficient RF amplifiers difficult. Both these pathologies cause deterioration of the system performance. Techniques to alleviate their effect on performance are described. Several standards that employ OFDM are provided, along with typical operational values that these systems use.

Finally, a system employing OFDM to transmit data modulated using binary phase shift keying (BPSK) and quadrature phase shift keying (QPSK) modulation is simulated. Monte-Carlo type simulations are employed to evaluate the performance of these systems in terms of the probability of error. Additionally, the effect of CFO on the error rate performance is illustrated. Lastly, the PAPR of an OFDM system is computed and shown to illustrate the fluctuations in transmit power. All of these simulations are performed in MATLAB. Programs used for these simulations are also provided.

# APPENDIX A

# Abbreviations

| | |
|---|---|
| A/D | Analog to Digital |
| AWGN | Additive White Gaussian Noise |
| BER | Bit Error Rate |
| BPSK | Binary Phase Shift Keying |
| CFO | Carrier Frequency Offset |
| CP | Cyclic Prefix |
| D/A | Digital to Analog |
| DAB | Digital Audio Broadcast |
| DFT | Discrete Fourier Transform |
| DMB-T | Digital Multimedia Broadcasting-Terrestrial |
| DVB-T | Digital Video Broadcasting-Terrestrial |
| FDM | Frequency Division Multiplexing |
| FFT | Fast Fourier Transform |
| FIR | Finite Impulse Response |
| HiperLAN | High Performance LAN |
| IBI | Inter-Block-Interference |
| IEEE | Institute of Electrical and Electronics Engineers |
| IFFT | Inverse Fast Fourier Transform |
| ISDB | Integrated Services Digital Broadcasting |
| ISI | Inter-Symbol-Interference |
| LAN | Local Area Network |
| LOS | Line of Sight |
| MAN | Metropolitan Area Network |
| MC | Multi-Carrier |
| OFDM | Orthogonal Frequency Division Multiplexing |
| PAPR | Peak to Average Power Ratio |

| | |
|---|---|
| PDF | Probability Distribution Function |
| QAM | Quadrature Amplitude Modulation |
| QPSK | Quadrature Phase Shift Keying |
| RMS | Root Mean Square |
| SC | Subcarrier |
| SNR | Signal to Noise Ratio |
| WLAN | Wireless Local Area Network |
| Wi-Fi | Wireless Fidelity Alliance |
| ZP | Zero Padded |

# APPENDIX B

# Notations

| | |
|---|---|
| $\geq$ | Greater than or equal to |
| $\leq$ | Less than or equal to |
| $\gg$ | Much greater than |
| $\ll$ | Much smaller than |
| $:=$ | Is defined as |
| $E[\cdot]$ | Expectation operator |
| $I_n(\cdot)$ | $n^{th}$ order Modified Bessel function |
| $\lvert x \rvert$ | Absolute value |
| $*$ | Linear Convolution |
| $[\cdot]^T$ | Transpose operator |
| $[\cdot]^H$ | Hermitian or the conjugate transpose operator |
| $\mathbf{F}_N$ | $N \times N$ DFT matrix |
| $\mathbf{R}_{cp}$ | Matrix used for removing cyclic prefix |
| $\mathbf{T}_{cp}$ | Matrix used for adding cyclic prefix |
| $\Delta f$ | Frequency offset |
| $\delta[\cdot]$ | Kronecker delta function |
| $\mathbf{T}_{zp}$ | Matrix to zero pad |
| $\mathbf{I}_N$ | Identity matrix of dimension $N \times N$ |
| $\mathbf{0}_{M \times N}$ | Zero Matrix of dimension $M \times N$ |
| $\tau_{max}$ | Maximum path delay |
| $\tau_{avg}$ | Average path delay |
| $\tau_{rms}$ | RMS path delay |
| $S(\tau, f)$ | Scattering function with delay $\tau$ and frequency $f$ as parameters |

# Bibliography

[1] E. B. Union, "ETSI TS 102 563 v1.1.1 - DAB+ enhancement specification," available online at http://www.etsi.org/; last accessed 07 Jan, 2010., 2007. 1, 21

[2] T. D. V. B. Project, "EN 302 755 V1.1.1 - Frame structure channel coding and modulation for a second generation digital terrestrial television broadcasting system (DVB-T2)," available online at http://www.etsi.org/; last accessed 07 Jan, 2010., 2009. DOI: 10.1109/18.850663 1, 21

[3] WorldDMB, "ETSI T-DMB TS 102 427," available online at http://www.etsi.org/; last accessed 07 Jan, 2010., 2005. 1, 21

[4] M. Doelz, E. Heald, and D. Martin, "Binary data transmission techniques for linear systems," *Proceedings of the IRE*, vol. 45, no. 5, pp. 656–661, 1957. 1

[5] G. A. Franco and G. Lachs, "An orthogonal coding technique for communications," *IRE International Convention Record*, vol. 9, pp. 126–133, 1961. DOI: 10.1109/TCOM.1964.1088883 2

[6] R. W. Chang, "Synthesis of band-limited orthogonal signals for multichannel data transmission," *Bell Sys. Techn. Journal*, vol. 45, pp. 1775–1796, Dec 1966. 2, 20

[7] P. A. Bello, "Selective fading limitations of the KATHRYN modem and some system design considerations," vol. 13, pp. 320–333, Sep 1965. DOI: 10.1016/0005-1098(91)90003-K 2

[8] M. S. Zimmerman and A. L. Kirsch, "The AN/GSC-10 (KATHRYN) variable rate data modem for HF radio," vol. 15, pp. 197–203, 1967. 2

[9] E. N. Powers and M. S. Zimmermann, "A digital implementation of a multichannel data modem," in *Proc. IEEE Int. Conf. Commun.*, Philadelhphia, PA, 1968. DOI: 10.1109/CDC.2005.1582620 2

[10] R. W. Chang and R. A. Gibby, "A theoretical study of performance of an orthogonal multiplexing data transmission scheme," vol. 16, pp. 529–540, Aug. 1968. 2

[11] B. Saltzberg, "Performance of an efficient parallel data transmission system," vol. 15, pp. 805–811, Dec. 1967. 2, 20

## 62 BIBLIOGRAPHY

[12] S. B. Weinstein and P. M. Ebert, "Data transmission by frequency-division multiplexing using the discrete Fourier transform," vol. 19, pp. 628–634, 1971. DOI: 10.1016/0169-7552(89)90019-6 2

[13] B. Hirosaki, "An orthogonally multiplexed QAM system using the discrete Fourier transform," vol. 29, no. 7, pp. 982–989, Jul 1981. 2

[14] G. Stuber, J. Barry, S. Mclaughlin, Y. Li, M. Ingram, and T. Pratt, "Broadband MIMO-OFDM wireless communications," *Proceedings of the IEEE*, vol. 92, no. 2, pp. 271–294, 2004. DOI: 10.1007/BF01386390 2

[15] M. Engels, *Wireless OFDM Systems: How to make them work?* Kluwer Academic Pub, 2002. DOI: 10.1109/MCOM.2009.4785383 2

[16] K. Fazel and S. Kaiser, *Multi-carrier and spread spectrum systems.* Wiley, 2003. DOI: 10.1109/25.260747 2

[17] A. Goldsmith, *Wireless Communications*, 1st ed. New York: Cambridge University Press, 2005. 5, 38

[18] A. Papoulis and S. U. Pillai, *Probability, Random Variables and Stochastic Processes with Errata Sheet*, 4th ed. McGraw-Hill Science/Engineering/Math, Dec. 2001. 6

[19] J. Proakis, *Digital Communications.* Mc Graw Hill, 4th Edition, 2001. DOI: 10.1145/316194.316231 7, 37, 38

[20] T. M. Duman and A. Ghrayeb, *Coding for MIMO Communication Systems.* New York: John Wiley and Sons, 2007. 7, 11

[21] W. Braun and U. Dersch, "A physical mobile radio channel model," vol. 40, no. 2, pp. 472–482, May 1991. 9

[22] A. Paulraj, R. Nabar, and D. Gore, *Introduction to space-time wireless communication.* The Pitt building, Trumpington street, Cambridge, United Kingdom: University of Cambridge, 2003. DOI: 10.1145/52324.52356 9

[23] COST 207, "Digital land mobile radio communications," *Office for official publications of the European communities, Final Report, Luxembourg*, 1989. 10

[24] J. Proakis, *Digital Communications.* Mc Graw Hill, 4th Edition, 2001. 11, 12

[25] H. Sari, G. Karam, and I. Jeanclaude, "Transmission techniques for digital terrestrial TV broadcasting," *IEEE Communications Magazine*, vol. 33, no. 2, pp. 100–109, February 1995. 12, 21

[26] T. Söderström and P. Stoica, *System Identification.* Prentice Hall, 1989. 12, 27

[27] M. Speth, S. Fechtel, G. Fock, and H. Meyr, "Optimum receiver design for wireless broad-band systems using OFDM: Part I," *IEEE Transactions on Communications*, vol. 47, no. 11, pp. 1668–1677, 1999. DOI: 10.1145/1151659.1159942 15

[28] R. Prasad, *OFDM wireless multimedia communications*. Boston, London: Artech House, 2000. 16, 34

[29] Z. Wang and G. B. Giannakis, "Wireless multicarrier communications: where Fourier meets Shannon," vol. 17, no. 3, pp. 29–48, May 2000. 16, 18, 20

[30] ——, "Linearly precoded or coded OFDM against wireless channel fades?" 2001, pp. 267–270. DOI: 10.1109/90.811451 17

[31] G. B. Giannakis, "Filterbanks for blind channel identification and equalization," vol. 4, pp. 184–187, June 1997. DOI: 10.1287/opre.9.3.383 18

[32] S. B. Weinstein and P. M. Ebert, "Data transmission by frequency-division multiplexing using the discrete Fourier transform," vol. 19, pp. 628–634, 1971. 18

[33] J. A. C. Bingham, "Multicarrier modulation for data transmission: an idea whose time has come," vol. 28, no. 5, pp. 5–14, May 1990. 18, 20

[34] T. Keller and L. Hanzo, "Adaptive multi-carrier modulation: A convenient framework for time-frequency processing in wireless communications," vol. 88, no. 5, pp. 611–640, May 2000. DOI: 10.1109/90.879343 18, 23

[35] H. F. Harmuth, "On the transmission of information by orthogonal time functions," *AIEE. Trans. (Commun. Electron.)*, vol. 79, pp. 248–255, 1960. DOI: 10.1109/MVT.2006.307304 20

[36] R. Negi and J. Cioffi, "Pilot tone selection for channel estimation in a mobile OFDM system," *IEEE Transactions on Consumer Electronics*, vol. 44, no. 3, pp. 1122–1128, 1998. DOI: 10.1109/TCOM.1987.1096782 21

[37] 802.11 Working Group, "IEEE Standard 802.11a-1999 - Higher Speed PHY Extension in the 5GHz Band," available online at http://standards.ieee.org/; last accessed 07 Jan, 2010., 1999. DOI: 10.1109/TNET.2007.900405 21, 26, 27

[38] S. Brandes, I. Cosovic, and M. Schnell, "Sidelobe suppression in OFDM systems by insertion of cancellation carriers," in *2005 IEEE 62nd Vehicular Technology Conference, 2005. VTC-2005-Fall*, vol. 1, 2005. 23

[39] ——, "Reduction of out-of-band radiation in OFDM systems by insertion of cancellation carriers," *IEEE Communications Letters*, vol. 10, no. 6, pp. 420–422, 2006. DOI: 10.1109/90.234856 23

[40] P. Robertson and S. Kaiser, "The effects of Doppler spreads in OFDM (A) mobile radio systems," in *IEEE VTS 50th Vehicular Technology Conference, 1999. VTC 1999-Fall*, vol. 1, 1999. 23

[41] P. H. Moose, "A technique for orthogonal frequency division multiplexing frequency offset correction," *IEEE Transactions on Communications*, vol. 42, no. 10, pp. 2908–2914, 1994. 25, 27

[42] T. Schmidl and D. Cox, "Robust frequency and timing synchronization for OFDM," *IEEE Transactions on Communications*, vol. 45, no. 12, pp. 1613–1621, 1997. 25, 27

[43] J. Armstrong *et al.*, "Analysis of new and existing methods of reducing intercarrier interference due to carrier frequency offset in OFDM," *IEEE Transactions on Communications*, vol. 47, no. 3, pp. 365–369, 1999. DOI: 10.1145/357401.357402 25, 29

[44] T. Chiueh and P. Tsai, *OFDM Baseband Receiver Design for Wireless Communications*. Wiley, Dec. 2007. DOI: 10.1145/584091.584093 26

[45] Y. Lim and J. Lee, "An Efficient Carrier Frequency Offset Estimation Scheme for OFDM System," in *IEEE Vehicular Technology Conference*, vol. 5. IEEE; 1999, 2000, pp. 2453–2458. DOI: 10.1109/TIT.2006.874390 27

[46] M. Morelli and U. Mengali, "An improved frequency offset estimator for OFDM applications," *IEEE Communications Letters*, vol. 3, no. 3, pp. 75–77, 1999. 27

[47] H. Minn, P. Tarasak, and V. Bhargava, "OFDM frequency offset estimation based on BLUE principle," in *IEEE Vehicular Technology Conference*, vol. 2, 2002, pp. 1230–1234. DOI: 10.1109/JSAC.2009.090207 27

[48] U. Tureli, H. Liu, and M. Zoltowski, "OFDM blind carrier offset estimation: ESPRIT," *IEEE Transactions on Communications*, vol. 48, no. 9, pp. 1459–1461, 2000. 27

[49] B. Chen and H. Wang, "Blind estimation of OFDM carrier frequency offset via over-sampling," *IEEE Transactions on Signal Processing*, vol. 52, no. 7, pp. 2047–2057, 2004. DOI: 10.1109/MCOM.2002.1018018 27

[50] M. Schmidl and D. Cox, "Blind synchronisation for OFDM," *Electronics Letters*, vol. 33, no. 2, pp. 113–114, 1997. DOI: 10.1109/9.182479 27

[51] H. Liu and U. Tureli, "A high-efficiency carrier estimator for OFDM communications," *IEEE Communications Letters*, vol. 2, no. 4, pp. 104–106, 1998. 27

[52] Y. Zhao and S.-G. Haggman, "Intercarrier interference self-cancellation scheme for OFDM mobile communication systems," *Communications, IEEE Transactions on*, vol. 49, no. 7, pp. 1185–1191, July 2001. 28

[53] ——, "Sensitivity to Doppler shift and carrier frequency errors in OFDM systems-the consequences and solutions," in *Vehicular Technology Conference, 1996. 'Mobile Technology for the Human Race'., IEEE 46th*, vol. 3, Apr-1 May 1996, pp. 1564–1568. 28

[54] M. Gudmundson and P.-O. Anderson, "Adjacent channel interference in an OFDM system," in *Vehicular Technology Conference, 1996. 'Mobile Technology for the Human Race'., IEEE 46th*, vol. 2, Apr-1 May 1996, pp. 918–922. 28

[55] C. Muschallik, "Improving an OFDM reception using an adaptive Nyquist windowing," *Consumer Electronics, IEEE Transactions on*, vol. 42, no. 3, pp. 259–269, Aug 1996. 28

[56] S. Weinstein and P. Ebert, "Data transmission by frequency-division multiplexing using the discrete Fourier transform," *Communication Technology, IEEE Transactions on*, vol. 19, no. 5, pp. 628–634, October 1971. 28

[57] J. Cimini, L., "Analysis and simulation of a digital mobile channel using orthogonal frequency division multiplexing," *Communications, IEEE Transactions on*, vol. 33, no. 7, pp. 665–675, Jul 1985. 28

[58] A. Spanias, *DSP: An Interactive Approach.*   North Carolina: Lulu Books, 2007. 28

[59] T. Wilkinson and A. Jones, "Minimisation of the peak to mean envelope power ratio of multi-carrier transmission schemes by block coding," in *Vehicular Technology Conference, 1995 IEEE 45th*, vol. 2, Jul 1995, pp. 825–829 vol.2. 34

[60] M. Golay, "Complementary series," *Information Theory, IRE Transactions on*, vol. 7, no. 2, pp. 82–87, April 1961. 34

[61] R. Sivaswamy, "Multiphase complementary codes," *Information Theory, IEEE Transactions on*, vol. 24, no. 5, pp. 546–552, Sep 1978. 34

[62] R. Frank, "Polyphase complementary codes," *Information Theory, IEEE Transactions on*, vol. 26, no. 6, pp. 641–647, Nov 1980. 34

[63] B. Popovic, "Synthesis of power efficient multitone signals with flat amplitude spectrum," *Communications, IEEE Transactions on*, vol. 39, no. 7, pp. 1031–1033, Jul 1991. 34

[64] S. Muller and J. Huber, "OFDM with reduced peak-to-average power ratio by optimum combination of partial transmit sequences," *Electronics Letters*, vol. 33, no. 5, pp. 368–369, Feb 1997. 34

[65] D. Tse and P. Viswanath, *Fundamentals of Wireless Communication.*   Cambridge University Press, June 2005. 37

[66] N. Metropolis and S. Ulam, "The Monte Carlo method," *Journal of the American Statistical Association*, vol. 44, no. 247, p. 335–341, 1949. 38

[67] R. E. Caflisch, *Monte Carlo and quasi–Monte Carlo methods*, ser. Acta Numerica.   Cambridge University Press, 1998, vol. 7. 38

[68] J. M. Hammersley and D. C. Handscomb, *Monte Carlo Methods*.   London: Methuen, 1975. 38

[69] C. P. Robert and G. Casella, *Monte Carlo Statistical Methods*, 2nd ed.   New York: Springer, 2004. 38

[70] S. M. Kay, *Fundamentals of Statistical Signal Processing: Estimation Theory*.   PTR Prentice-Hall, Inc., 1993. 38

[71] T. Pollet, M. Van Bladel, and M. Moeneclaey, "BER sensitivity of OFDM systems to carrier frequency offset and Wiener phase noise," *IEEE Transactions on Communications*, vol. 43, no. 234, pp. 191–193, 1995. 44

[72] K. Sathananthan and C. Tellambura, "Probability of error calculation of OFDM systems with frequency offset," *IEEE Transactions on communications*, vol. 49, no. 11, pp. 1884–1888, 2001. 44

# Authors' Biographies

## ADARSH NARASIMHAMURTHY

**Adarsh Narasimhamurthy** is a Ph.D. candidate at the School of Electrical, Computer and Energy Engineering in Arizona State University, Tempe. He obtained his B.E. with distinction in 2005 from the Bangalore University, Karnataka, India and the M.S. degree in Electrical Engineering from Arizona State University, Tempe in 2007. Currently, he is a part of the Signal Processing for Wireless Communications Lab headed by Dr. Cihan Tepedelenlioğlu and also a member of the SenSIP consortium.

 Mr. Narasimhamurthy was awarded a Research Assistantship in the year 2006 and the subsequent year a Graduate Teaching Associate position from the Department of Electrical Engineering. His research interests include MIMO systems, OFDM systems, reduced complexity diversity combining techniques and multiuser communication. Mr. Narasimhamurthy is also a member of the Eta Kappa Nu honor society and an IEEE student member.

## MAHESH K. BANAVAR

**Mahesh K. Banavar** received the B.E. degree in telecommunications engineering from Visvesvaraya Technological University, Karnataka, India, in 2005 and the M.S. degree in electrical engineering from Arizona State University, Tempe, in 2008. He is pursuing the Ph.D. degree with Arizona State University, specializing in Signal Processing and Communications, and doing research in wireless communications and sensor networks.

 Mr. Banavar is a member of the Eta Kappa Nu electrical and computer engineering honor society.

## CIHAN TEPEDELENLIOĞLU

**Cihan Tepedelenlioğlu** was born in Ankara, Turkey in 1973. He received his B.S. degree with highest honors from Florida Institute of Technology in 1995, and his M.S. degree from the University of Virginia in 1998, both in Electrical Engineering. From January 1999 to May 2001 he was a research assistant at the University of Minnesota, where he completed his Ph.D. degree in Electrical and Computer Engineering. He is currently an Associate Professor of Electrical Engineering at Arizona State University. He was awarded the NSF (early) Career grant in 2001, and has served as an Associate Editor for several IEEE Transactions including IEEE Transactions on Communications, and IEEE Signals Processing Letters.

His research interests include statistical signal processing, system identification, wireless communications, estimation and equalization algorithms for wireless systems, multi-antenna communications, filterbanks and multirate systems, OFDM, ultra-wideband systems, distributed detection and estimation.

Printed in the United States
by Baker & Taylor Publisher Services